PAUL BOCK

HEVER

HI-SPEED ELECTRIC VEHICLE ELEVATED ROADWAY

Copyright © 2010 Paul Bock
All rights reserved.

ISBN: 1451567847
ISBN-13: 9781451567847

HEVER

Table of Contents

Foreword i
Chapter 1:   Why Build a New Transportation System. 1
Chapter 2:   Early History of Transport Evolution. 9
Chapter 3:   Individuality 15
Chapter 4:   Miles per Gallon. 23
Chapter 5:   Trains, Trucks, and Airplanes. 29
Chapter 6:   Our Fifty-Plus-Year-Old Interstate System . . . 37
Chapter 7:   Great Large Obstructive Block (GLOB) 41
Chapter 8:   Information Sign, Stop Sign, Traffic Signal . . 47
Chapter 9:   Interstate Design and Implementation. 51
Chapter 10: The Interstate Champion. 57
Chapter 11: Driving the Interstate 63
Chapter 12: Human Actions While Driving 71
Chapter 13: The Future. 75
Chapter 14: Switches, Brakes, and Sensors 83
Chapter 15: Wheels, Motors, Spurs, and Parking 87
Chapter 16: Some HEVER System Benefits 91
Chapter 17: Testing 97
Chapter 18: The HEVER system 101
Chapter 19: Trucks 107
Chapter 20: Automobiles 117
Chapter 21: Buses (and Limos) 125
Chapter 22: Automobile and Truck Non-compatibility . . . 131
Chapter 23: Operating the HEVER system 135
Chapter 24: Driving On and Off the HEVER System 139

Chapter 25: Manufacture, Distribute - The Kings 145
Chapter 26: Simulation 149
Chapter 27: Solar Energy 153
Chapter 28: Electromagnetics 159
Chapter 29: Maglev Train Electromagnetic Levitation . . . 167
Chapter 30: Data Entry, Gathering, and Its Use 171
Chapter 31: Construction and Maintenance - HEVER . . . 177
Chapter 32: Inertial Forces 179
Chapter 33: Parking Vehicles Leaving HEVER 183
Chapter 34: Some Notes on Future and Change 187
Chapter 35: Organization for Design/Implementation . . . 191
Chapter 36: Computer Systems 197
Epilogue: Reasons for Writing a Book 201
Bibliography 207

Paul Bock Resides in Fallon Missouri

1964 BS Southern Illinois University
1965 CDP Certificate in Data Processing from DPMA
1995 MBA Lindenwood University

44 years in the Computer field - Programmer, Scientific Programmer, Systems Programmer, Operating Systems Programmer/Designer, IT Manager, Computer Store Company Owner, Consultant and Project Manager.

20 years - Adjunct (part time) Professor of Computer Programming Language Courses.

I dedicate this book to my wife Linda. She has patiently stood by and supported me no matter what.

Our families have been very supportive. With a special thanks to my daughter Michelle Hastings, she helped and encouraged me on publishing via the Internet.

I would like to acknowledge and thank my "Book Writing" Professor, Scottie Priesmeyer. Without her encouragement and sharing of knowledge, I could not have written this book.

Now is the time for all good men to come to the aid of our country.
One for all; all for one; bring everybody home. Is this just pleasantry?

Foreword

Hi-speed Electric Vehicle Elevated Roadway (HEVER)—people have told me that this idea is exactly what we need. We burn too much oil that we should save for use in the manufacturing of plastics, for production of fuels to use in defense of our country, and for space exploration. Virtually all nations produce too much pollution by burning toxic materials, especially oil. Some nations have just started to pollute, others have increased their usage of toxic materials. Global demand has increased significantly for toxic materials such as oil, much of which our country subsequently burns. We must provide more efficient use of any fuel and more development of electricity production, and especially more solar electricity production.

When the Interstate was new, its development included a national roadway design using available technology of that time (1950s). Having now reached its limit, the Interstate upgrades are no longer an overall system design; the system can only be patched. The new system must be designed, not patched; our current Interstate system can only allow small to large patches. Today we can't integrate roadway design changes into our current Interstate system.

The HEVER roadway will prove justifiable purely for commercial moving of products and material; however, the greatest benefit, saving lives, will occur due to the safe and hi-speed movement of people. We need faster speed to shorten the time for movement of products, materials, and people non-stop from their place of origination to their needed destinations.

In this book, I discuss travel at 400 mph. However, the real barrier to higher speeds is the sound barrier. These newly designed vehicles actually could travel at speeds greater than 400 mph, and even far greater than the speed of sound; however, crashing through the sound barrier might cause damage, such as broken windows. These newly designed vehicles could easily travel as fast as the speeds of the fastest non-overseas commercial jet airplanes. Except for the practical limit of ground transport not exceeding the sound barrier, electromagnetic propulsion, like that used by maglev trains, could propel vehicles far in excess of 1000 mph.

The HEVER system will be complementary to our existing Interstate system. This new system will actually help reduce maintenance cost and accidents on the Interstate. Safety on the Interstate highway will improve and road wear will reduce because fewer vehicles (especially large trucks) will travel on the Interstate. Except for safety reasons, this new transport system will make it unnecessary to expand or change the Interstate system.

The HEVER system will use electromagnetic propulsion like that used by maglev trains. Maglev, a term applied to a hi-speed type of train, means magnetic levitation. Electro-magnetic propulsion uses electro-magnets to levitate and hold maglev trains in place about a half inch to almost four inches above the surface, and causes the forward or backward movement of the train. This concept allows maglev trains to travel over 400 mph. Maglev trains travel at a speed faster than bullet trains and three or four times as fast as most conventional trains that have an older design.

In many cases, maglev trains will still have to make intermediate stops at stations to pick up and drop off travelers and materials. Most bullet and maglev trains will only reduce time to get to most destinations by about one half of what slower, more convention trains can do. For example, replacing the train used on a current Amtrak run, between St. Louis and Chicago, with a hi-speed train going 200 mph, the eight-hour trip would reduce to four hours. The distance between St. Louis and Chicago is a little over two hundred miles, yielding an average of a little over 50 mph. Due to the freedom of today's automobiles and trucks to go anywhere and on an almost non-stop basis, trains still would usually not compete with trucks and automobiles.

FOREWORD

Trucks, automobiles, and buses using the proposed HEVER system will have the ability to travel non-stop from the beginning of a trip to its end (as an option, absolutely no intermediate stops), thereby fully benefitting from the high speed allowed by electromagnetic propulsion like that used by maglev trains.

Maglev trains operating on a closed roadway using the otherwise conventional train concept somewhat similar to that which exists today still will not handle human traffic with the flexibility provided by future maglev trucks, automobiles, and buses. The concept of maglev trucks, automobiles, and buses is the same as maglev trains except that they can provide the most flexibility in transport of people and the movement of freight.

The proposed system will have an elevated roadway with as gentle slopes as possible. Designers must likewise straighten out curves. The footprints for the system's structures holding the roadway will require much less area than required for today's railroad track beds. The book proposes mostly using the Interstate roadway right of ways for this new system with the goal of complementing the Interstate. While the book does not address building roadways on highway right of ways other than those of the Interstate, implementing the HEVER system on some of other US and state highways should occur later. Also, controlling traffic by computers on city streets is not addressed in this book. However, some day it will have to happen and will likely employ computer controlled, elevated streets. This is a far more difficult design endeavor than the design and construction of the HEVER system; this book does not and cannot come close to accomplishing such a design.

The proposed system will have trucks, buses, and automobiles that will travel under complete computer control while on this new system. Statistics gathering, entertainment, and travel information will be highly accessible while using the new system. On long trips, travelers (including truck drivers) can sleep, read, watch TV, or play cards.

This book is intended to promote movement to a new highly automated system; it will still need accomplishment of feasibility studies, simulation studies, and the development of requirements, specifications, and design documents. There is no intention to limit design to any of the ideas in this book; if designs are present, they are intended as examples.

In our transportation systems, today, we often use one or two century-old transportation technologies. Some of the technology still provides good results and some of it has become outmoded. Much of the automotive industry products provide about as much safety and comfort as the industry can provide. One of the big problems is the high cost of fuel, which more people don't seem to mind when the country has a strong economy.

Future design of all-electric vehicles must, at a minimum, provide safety and comfort; availability of electricity will result in lower fuel cost. When using the roadway while traveling on the new system discussed in this book, this type of vehicle will be able to be transported as freight, and also operate as a conventional vehicle when not on the HEVER system. Past vehicle designs will not limit its future designs. Vast improvement in safety and comfort will happen because of more computer monitoring and control, and the newly designed vehicles will provide more electrical and electronic options. The future vehicles will solely use electric power as fuel.

From the start of this endeavor, the country will provide jobs in technology, engineering, project management, and construction, along with helping many second tier companies that would benefit from providing materials and components for construction and operation of this new system. Workers in the US can do most, if not all, of the work.

Vacation bound: save gas, save time, save gas, save time, shut the car door.
Everybody's in the car; step on the gas; let her roar.

Chapter 1:
Why Build a New Transportation System

A main concept in this book strongly presents the concept that we must design a new transportation system that will give us a significantly improved way of traveling and transporting. Ideas about these new processes and new systems can only find implementation if the ideas are valid and if someone in authority is willing to accept risk. Ideas only have a chance to come to fruition if we reasonably understand them. Accomplishment of this understanding can happen through the choice and employment of critical thinkers. They usually have a higher intelligence and more experience about some choices for change than persons in authority who would have to make decisions and tolerate the risk. These persons in top executive positions are generally very intelligent. They usually have much more business savvy than most technically intelligent persons, but they often can't understand and do not find it easy to communicate with technically smart persons. The persons in authority, especially executives and high-level government officials, must make enough of an effort to understand these intelligent thinkers. Some level of risk always goes along with the approval and implementation of change. If enough due diligence is done prior to executing the change, the results can prove gratifying.

An article, "Why Design Thinking Matters," written by Helen Walters, appeared on the *BusinessWeek* internet site. A consultant to top executives during much of his past career, and now dean of the School of Management at the *University of Toronto*, Roger Martin has given top executives of companies like Procter

HEVER

& Gamble and Research In Motion expertise that has helped them considerably. He contends that companies faced with good sounding ideas that would require potential for significant risk/reward, avoid approving the making of changes because persons in authority often lack a risk aversion capability. Some projects don't even get considered because top executives quite often don't want to be the first in their company or industry to try some new idea. They don't want to be pioneers.[1]

People, facing potential change that new projects can bring, should not fear the necessity for change because they have some uncomfortable feelings with some issues. Rather, an in-depth, professionally engineered judgment usually will yield the best results.

Much planning and design will be required to implement a new computer controlled system that can complement the Interstate. A great amount of planning and design went into the creation of the existing Interstate system. However, having a large, relatively old Interstate system makes it costly to maintain, and especially difficult and expensive to upgrade. We currently can only do limited maintenance and upgrades because they must fit into our present Interstate system. This limits any range of change, small or large. The original designers of our current Interstate system determined the system requirements and its design. Designers limited the ability to easily expand the Interstate by choosing the types of overpasses with roadbed areas often too narrow to allow addition of more lanes. They also limited the ability to easily expand by not building the right overpass accesses to the Interstate. Also road planners often did not procure enough property for additional lanes when it was obvious that as the population continues to grow the roadway will need more lanes. We call this extensibility, the ability, or lack thereof, to expand due to original design. But, in the future, our country will obviously need good roadway accesses.

The current Interstate system has lost its extensibility. The system can only experience fixes, patches, addition of lanes, and make a few other types of changes; an option of total overall design no longer exists. Constriction caused by construction often causes

---
[1] Helen Waters, "Why Design Thinking Matters," October 27, 2009, http://www.businessweek.com.

WHY BUILD A NEW TRANSPORTATION SYSTEM

closures of roadways, delaying motorists. This sometimes causes temporary and lengthy closures. The U.S. Bureau of Transportation Statistics reported the following 2006 U.S. vehicle statistics.

Passenger car total	135,399,945
Vans, pickups, and SUVs total	99,124,775
Truck total	8,819,007
Bus total	83,902
Class 1 freight car total	475,415
Class I locomotives total	23,732
Non-class I freight car total	120,688
Light and heavy rail car total	12,853
Light and commuter rail car and locomotive total	6,403
Amtrak passenger train car total	1,191
Amtrak locomotive total	319[2]
System highway mileage	4,032,126
System Class I rail mileage	94,440
Class I Amtrak mileage	21,708[3]

Our society has not designed or planned new, integrated alternatives to our existing Interstate road systems for use by trucks, automobiles, and buses; we have been modifying our Interstate and other roadways for over fifty years through patches. Our population has grown in the past fifty years. Due to births, people living longer, and immigration, the population will grow even more in the next fifty or a hundred years. Population growth causes increased traffic volumes. Constantly increasing traffic volume has overloaded our road systems. The number of accidents increases with time. We waste oil-based fuel that will someday run out. Use of this fossil fuel continues to add pollution to our environment.

Navigating Interstate roadways has an inherent probability for the occurrence of accidents. This accident probability is exacerbated by our increasingly overloaded roadways. We drive on them using quite inefficient vehicles controlled by humans with

---

[2]   U.S. Bureau of Transportation Statistics, http://www.census.gov/compendia/statab/tables/09s1027.xls.
[3]   U.S. Bureau of Transportation Statistics, http://www.bts.gov/publications/national_transportation_statistics/html/table_01_01.html.

varying driving abilities. The operation of these vehicles pollutes the air, wastes ground space, burns an irreplaceable resource that we shouldn't waste, and often encourages drivers of vehicles to become extremely aggressive. Usually aggressive actions occur because drivers feel that they must aggressively compete or that other drivers will merely take advantage of them, for example, by cutting in front of them. This rationale, causing the aggression, exists for no good reason; yet as our option, we often decide to make it the thing to do.

The U.S. Department of Transportation (US DOT) reported the following, U.S. vehicle annual fuel usage statistics.

Barrels per day, U.S. petroleum consumption	20,690,000[4]
Gallons wasted annually due to congestion	2,514,400,000[5]
Gallons wasted fuel annually by drivers	1,795,000[6]
Gallons used annually by class I freight service	4,192,000,000[7]

Currently, when making roadway construction changes, called patches (not an overall integrated design), they often cause delay to a relatively large numbers of drivers of automobiles and trucks. Because of this ongoing construction, sometimes the need to re-route some of these vehicles happens, sometimes for long periods of time. Accident and gaper blocks cost individuals needing to use the Interstate valuable time and sometimes money; emergency personnel quite often find it difficult to respond to traffic and accident problems.

---

4 *U.S. Department of Transportation (US DOT), http://www.bts.gov/publications/national_transportation_statistics/2008/excel/table_04_01.xls.*
5 *U.S. Department of Transportation (US DOT), http://www.bts.gov/publications/national_transportation_statistics/2008/excel/table_04_27.xls*
6 *U.S. Department of Transportation (US DOT), http://www.bts.gov/publications/national_transportation_statistics/excel/table_04_28.xls.*
7 *U.S. Department of Transportation (US DOT), http://www.bts.gov/publications/national_transportation_statistics/2008/excel/table_04_17.xls.*

WHY BUILD A NEW TRANSPORTATION SYSTEM

YEAR	1993	1994	1995	1996	1997	FATALITIES	INJURIES
Fatalities	42,827	43,587	44,578	44,848	44,474	220,314	
Injuries	3,223,298	3,345,263	3,539,389	3,554,305	3,417,846		17,080,101

YEAR	1998	1999	2000	2001	2002	FATALITIES	INJURIES
Fatalities	43,910	44,084	44,384	44,941	45,297	222,616	
Injuries	3,262,309	3,305,649	3,259,673	3,100,080	3,958,911		15,886,622

YEAR	2003	2004	2005	2006	2008	FATALITIES	INJURIES
Fatalities	45,101	44,985	45,565	44,974	43032	223,657	
Injuries	2,918,528	2,818,446	2,728,327	2,604,648			11,069,949

Total Number of Fatalities	666,587
Total Number of Injuries	44,036,672

Every year, 40,000 people die on our highways, over 660,000 more are injured. We can and must virtually eliminate deaths and injury due to accidents on our highways. Each year way too many people die or are injured in traffic accidents.

From 1993 through 2006, over 44,000,000 vehicle accidents were reported (injuries were not reported in 2007), and from 1993 through 2007 a total of over 666,000 persons died.[8]

Imperfect drivers, who are sometimes quite poor drivers, navigate our roads. They sometimes act discourteously by shouting obscenities or they make threatening gestures at other drivers. Sometimes they even become violent (road rage). Often they involve themselves in breaking traffic laws, such as speeding or ignoring yield signs, stop signs, or red lights. They sometimes go the wrong direction on a one-way street or make a left turn at a no left turn intersection, and some even drive under the influence of alcohol or drugs. Many of these violations cause dangerous and sometimes deadly situations.

---

[8] U.S. Department of Transportation, http://www.bts.gov/publications/national_transportation_statistics/pdf/entire.pdf.

HEVER

A front page article, "Drivers Ignore Hwy. 40 Barriers," written by Elisa Crouch, appeared in the *St Louis Post Dispatch* about drivers who made poor choices ignoring Highway 40 barriers. The article illustrates how drivers often brazenly ignore common sense and do not avoid the obviously dangerous construction situations. The article stated that drivers would drive through in groups of automobiles as construction workers watched, not able to stop them. The article seemed to imply that they drove through more as a thrill than confusion. Some gaps in the roadway under construction had a deep drop off of a foot or more. One motorist, who was going 75 mph, was stopped by police; he indicated to the police that he and his girlfriend were heading to an entertainment event and were using the unfinished roadway hoping to save time. The police stopped him just before he reached a massive hole where workers had recently removed an overpass. During construction of the HEVER system the roadway will be elevated and use by vehicles will not be possible. The main point of the article is to show how people so often make poor driving decisions.[9]

There is a high appetite for large automobiles and SUVs. An article from *Forbes* stated that, for the month of August through half of October, 2008 the sales of gas guzzling automobiles, trucks, and SUVs showed that buyers had a high propensity for them. Of note in the article, the F-Series Ford truck and Chevrolet Suburban sales increased more than 3 percent. Sales of some big SUVs increased as much as 23 percent. During the month prior to this article, gas prices dropped over a dollar per gallon. The sudden significant dip in gas prices caused much of these increased sales figures. The sales of gas guzzlers will climb significantly as the economy continues to show more recovery. Many buyers of today's vehicles would prefer expensive, large, comfortable vehicles as long as they perceive that gas prices are not too high. After the economy improves enough, gas prices will increase and take back the dollar that was given when the economy bottomed. The gas prices will probably even go higher than merely taking back the

---

[9] Elisa Crouch, "Drivers Ignore Hwy. 40 Barriers," St. Louis Post-Dispatch, September 2, 2009.

WHY BUILD A NEW TRANSPORTATION SYSTEM

dollar price drop. The demand for SUVs and their resale values will again drop significantly.[10]

A new transportation system that is completely computer controlled will virtually eliminate these problems. The HEVER system will address mass transport of vehicles containing freight or people across the country. The word mass here relates to non-stop, controlled, hi-speed transport of a high volume of people and truck freight. The people in America need a safer, more efficient mass transportation system than what we have today, and one that operates at higher speed. Even for the current Interstate system, the new system would reduce maintenance costs while improving safety (e.g., fewer large trucks on the Interstate). Our country needs a new, safe, viable, extensible, and maintainable transportation system. This automated transportation system, the hallmark of this book, will usher in a new mass transit system and today's technology will accomplish the challenging job.

---

10   "Some Gas-guzzling Vehicles Enjoy Sales Surge," *Forbes*, October 23, 2009, http://www.msnbc.msn.com.

*London had smog, Pittsburgh had smog, Los Angeles has smog.
So many reasons, so many places; why try to change;
why is everyone so agog?*

Chapter 2:
Early History of Transport Evolution

Man wants to enjoy an increasingly better standard of living. The desire for man to act as an individual drives the central concept of this book. We must continue to increase our standard of living while providing faster and safer transportation. At the same time, we must eliminate most of the things that pollute our planet. Sometimes we have little or no choice because of the ubiquitous nature of the products produced by our existing institutions, such as the truck, automobile, and oil industries. We depend heavily on trucks and automobiles, which use oil in the form of gasoline, burned by the combustion engine. This represents the mode of transportation for the last one hundred years.

An amazing place, this planet Earth seems almost to have started from nothing and from a seemingly infinite environment. The concept of infinity seems to contradict the concept of time. If one believes that the environment's past existence goes backward in time forever (infinity), how can we ever live in any time frame, today, tomorrow or the future? The concept of the existence of infinity and also the concept of time can and do both exist relatively concurrently; we just don't know how. Most people believe that infinity exists now and that it always will. We don't know how long time will last. There exists a common perception is that it will not last forever, and yet still some other people believe in a possibility that the universe and, therefore, time, might last forever. We won't need time after life ends. Assuming that infinity exists, its source of existence must come from a dimension other than time.

One can think of time as a relational measurement of the Earth's rotation about its axis and its orbit around the Sun. This also implies that time measurements for other planets will differ from the Earth's time. Other planets have a different rotation than that of the Earth and different orbit sizes around Suns.

In an article "Oldest Human Skeleton Offers New Clues to Evolution," Mr. Azadeh Ansari a science writer for *CNN*, gives us an idea as to how long humans and their hominid ancestors have been evolving on the Earth. The oldest human remains, an ancient female, discovered in Ethiopia, lived on Earth approximately 4.4 million years ago. Actually, it is commonly believed that humans have evolved over a six million year period of time. Compared to the age of the Earth, humans have inhabited the Earth for such an almost unbelievably short period of time; yet, humans have made so many changes to the Earth. Furthermore, most of the significant changes that man has made have occurred in the past two hundred or so years; many of these changes have occurred to the detriment of the Earth. Hopefully the Earth will not actually wear out.[11]

The Earth evolved over a long period of time. Man has existed on Earth only for a relatively short period of time. During that time, while living basically as a social group, humans evolved from knowing very little to a point where they have gained significant knowledge. As society has applied this increasing knowledge, it sometimes has used it for group and individual gain. Due to an increase in knowledge, imagination, and invention, society has generally enjoyed an increasing standard of living. Quite often changes causing an increased standard of living have contributed to the inordinate and wasteful consumption of irreplaceable resources and added to pollution of the Earth's environment. Through all of this time, there persists one prevalent social characteristic, man's desire to accomplish things as an individual. That does not imply that man does not act in groups. Man often contributes his individual skills and knowledge to groups, teams, cities, and countries. However, man's desire for individual action and accomplishment causes pluses and minuses in the evolutionary development of our world.

---

11   Azadeh Ansari, "In the Blink of an Eye," CNN, October 7, 2009, http://www.cnn.com.

EARLY HISTORY OF TRANSPORT EVOLUTION

We now have so many established, entrenched institutions. Common examples include companies or industries that have created institutions in telephones, automobiles, oil production, Internet search, computers, and computer operating systems. To get an understanding of the great difficulty to make changes to our larger institutions, consider an analogy of changing the direction of large sea-going ships. These ships do not change direction quickly. Neither do our entrenched institutions.

Historically, many problems have faced society. England has had smog problems caused by burning coal even before the automobile was commonplace. Prior to 1960, and during especially cold winters, London pollution was the greatest; it was like a very thick stinky fog. Due to not using technology for cleaning smoke stacks (not using scrubbers), Pittsburgh also had a dirty air (smog?) problem from using coal in the steel mills. Workers in offices often carried an extra shirt to work, which they changed into in the afternoon. Man sometimes causes smog, like the situation in Los Angeles where, on humid days, a vast number of automobiles produce an inversion that creates a large cloud that virtually covers the entire city. In some other cases, natural causes such as volcanic action cause smog. However, the occurrence of most environments polluted by humans happens with far more regularity than those caused by volcano eruptions. Examples of pollution caused by humans includes plastic thrown into rivers and streams, oil or gas that car owners discard on open ground or put down into our sewers, gas spilled on lakes and rivers by boaters, burning of coal to heat homes or to produce electric energy for industry, and improperly discarding of radioactive hazardous waste.

Since the discovery of oil, no fuel has competed effectively with gas, which comes relatively easily and cheaply from the ground. Gas prices kept rising. Wanting to lower fuel costs, Brazil tried to have all of the country's vehicles operate 100 percent on alcohol. Automobile companies' vehicle production made considerable headway by manufacturing vehicles with engines that solely used alcohol for fuel. The venture failed because the oil industry merely lowered its prices. Cheaper prices do not always yield a safe enough situation or better long-term ownership cost.

HEVER

Along with transport improvements, came increased risk of damage or personal harm due to increases in speed and in height above the ground. While not always true, as a general rule, the slower that man walks or runs and/or the closer a man operates to the ground's surface, the lower the probability that he will experience the risk of damage or harm.

The horse and buggy era caused the implementation of the first roads, which originally served commercial needs. In theory, the roads would exist to move freight. Soon after, people deemed it necessary to use wagons and eventually buggies to carry passengers. The development of open buggies and other enclosed buggies (stagecoaches) soon followed.

When railroads came along, again, our forefathers originally intended to use them for the commercial moving of freight. People began hitching free rides on freight trains by merely jumping on when a train stopped or moved slow enough; trains initially did not run fast, two to five mph, certainly not much faster. Expansion of freight train capabilities soon included the addition of passenger cars and ticket sales.

For the most part, trucks greatly reduced much of the need for railroads, very similar to the way that railroads replaced the need for canal development. While design and development of high-speed bullet trains and maglev trains have appeared, they will likely not take away much business away from trucks. Proponents of this technology have proffered that this new development of high-speed railroad transport will reinvigorate the railroad industry and will retake its lost market. This will prove unlikely because trains must still make many stops and must often combine transport of people, products, and material in order to economically compete with trucks, automobiles, and buses.

The advent of the HEVER system will only exacerbate the difficulty for bullet trains and maglev trains to compete. Most of the market for trains will continue to involve the transport of low value, high volume material. Trains might not even retain all of this niche market. Keep in mind that maglev trucks, using the HEVER system, can offer a lower transport cost than today's trucks can, and they can deliver non-stop, one time loading and one time unloading of products and material from origin to destination.

Do we live in the past, was the middle ages so bad in the past?
Ah, I have a car and some gas; let's get there alone and fast, free at last.

Chapter 3:
Individuality

Man, at his option and as a free individual, will increasingly need the ability to act independently or as an individual who works or plays as part of a group.

As one of its main issues, this book looks at individuality as it relates to future needs of any proposed new mass transit system. Critical thinking has its roots in individuality and critical thinkers will perform well in helping to design future computer controlled roadways. In America, thinking as individuals has resulted in a nation normally leading the way rather than following. Individuals need to jealously guard their right and responsibility to think and act as individuals. When individuals become part of a group or team, they must still bring to the team their individual knowledge, experience, and thinking. When these groups exist as teams, individuals must act as individuals and also join the team; they must provide meaningful support as a teammate. In America, we often cater to young people and sometimes to the individual. Perhaps we should encourage more training, education, and guidance. These young people are our country's future.

According to MapQuest Maps, persons traveling the distance between St. Louis and Kansas City, Missouri in an automobile must drive 234 miles, and between St. Louis and Chicago, 260 miles. If we could safely drive our vehicles 250 mph, it would take approximately one hour to make a trip between these cities. Vehicles on the HEVER system will travel at 400 mph. At such a high speed, travelers will accomplish trips quickly. At 400 mph, it would take slightly over a half hour to make a trip between St. Louis and

HEVER

Kansas City. If necessary, or as an option, St. Louis workers could actually commute daily from St Louis to Kansas City or Chicago.

Driving your own electric vehicle on this new automated HEVER system would mean that you wouldn't have to experience the cost of leaving your car in a parking lot near the airport and waste time taking a shuttle to the airport. You would not have to get into a baggage check-in line and pass through an associated security line. You wouldn't have to arrive at the airport an hour or two before your flight takes off, hoping that the airplane will actually take off without the schedule delays that so often happen—there may also be a flight delay while the pilot waits for the control tower to let the plane taxi and take off.

You wouldn't have to experience the delay of your airplane landing at an airport as an intermediate stop (not your final destination)—the airplane taxiing to its required gate and allowing passengers on or off. If you must change planes, you will likely need to go through the terminal and determine which connecting flight you must use. Sometimes the amount of time to reach your connecting flight leaves you hoping your airplane has not left without you. If your plane arrived and landed late enough, you might have already missed your connecting flight. In that case, you would have to reschedule and perhaps you might have to use yet a different gate. Eventually you will find yourself on the correct airplane; it taxis to its assigned runway and takes off again.

You don't have to experience contacting a ticket agency or using the Internet to buy a ticket (taking the trip sometimes includes a luggage and tax surcharge). You don't have to waste time getting a boarding pass, or feel the pain of paying the high ticket price you pay for the sometimes cramped seating quarters.

You don't have to walk long distances in the airport to retrieve your luggage. You don't have to take a taxi or complimentary hotel limousine to the place where you are going to stay. Alternately, you don't have to rent a vehicle. You don't have to lug your baggage while you go to a place where you can pick up your rented automobile, which sometimes you must pick up from a remote location from the airport. You would likely have a similar experience, with some differences, on your return trip home.

INDIVIDUALITY

An article "U.S. Limits Waits on Planes" appeared on the Tuesday before Christmas 2009 in the St. Louis Post-Dispatch. It discussed the long waits passengers have to endure on taxiways. Transportation Secretary Ray LaHood ordered an airline to let passengers off an airplane; the passengers had been on the airplane over three hours. A new rule issued by Transportation Secretary Ray LaHood, to go into effect in 120 days, states that passengers must have food and water and a working bathroom, and, if needed, they would get medical attention within two hours of the plane delay occurring on a tarmac. If airlines violate a three-hour delay limit, the airline will be fined $27,500 per passenger. James May President and CEO of the Air Transport Association of America has stated that the airline industry will comply with the new ruling. Some airlines have reacted negatively, stating that this will make things worse, leading to unintended consequences such as cancelled flights and greater passenger inconveniences.[12]

At any speed, traveling on the HEVER system will result in a safer, more efficient and more enjoyable experience than flying. The maglev automobile will travel a fraction of an inch to a little less than four inches above its roadway, depending on the electromagnetic levitation method and devices used. Maglev vehicles will have virtually no friction drag like that caused by tires, and passengers will experience a very quiet ride. Levitating means that maglev vehicles will travel at a constant height above the roadway. The system will hold the vehicles by using magnets that exist on the roadway and the vehicle. An airplane has no vertical constraints that fix its distance from the ground.

In the future, most freight and humans will move at high speed using electric trucks and automobiles. Now we have a greater population than when trains enjoyed the height of their heyday. Can we prove whether high speed transporting of freight creates more value or if high-speed transport of humans creates more value? Because prevention of loss of human life ranks number one, high speed, safe transport of humans will win.

Trucks, buses, and automobiles in the future will more safely transport people and freight on an elevated roadway that will use

---

12 "U.S. Limits Waits on Planes," St Louis Post-Dispatch, December 12, 2009.

highly controlling computer systems. Most certainly, design and development of new transportation systems will include the automation of high-speed transport. This change will accomplish the automation of all single vehicle traffic, while providing a lower probability of harm to humans or damage to products and material.

Horses, wagons, buggies, sailboats, steamboats, railroads, airplanes, automobiles, buses, and motorcycles have all enjoyed eras of popularity. However, currently, society clings to an infatuation for the automobile (and trucks) that makes the automobile the preferred mode of human transport. When we design new, future systems, we must not forget man's need to act as an individual, especially as it relates to human transport systems.

In looking at the issue of human and freight transport, one can look at expansion of the existing Interstate system as the way to handle increasing traffic volumes as the population increases. As the population increases, more people will drive and, therefore, this growth eventually will impact the situation and cause the perception that we need to expand streets and expressways. More people flying causes the need for more flights and the eventual consideration for a perceived need for the expansion of airports.

Some traffic relief might occur because of an increase of train speed, more efficient use of automobiles and trucks, and additional light rail. Our society enjoys better transportation institutions than that which our ancestors did, but now our transportation systems suffers from neglect, overload, and old age. If one observes vehicles on streets and expressways, one will most often see automobiles and trucks that only have one occupant. As a direct result, man's need and/or desire for individual action to accomplish his personal goals and objectives, causes this. This need for individuality won't lessen in the future. This gives us yet another reason to eliminate the use of oil and use electricity, the more efficient and lower cost fuel.

On an average, airplanes, trains, ships, and buses will normally transport a larger number of passengers than individual automobiles and trucks. Traveling in automobiles and trucks rather than airplanes or buses usually results in a lower probability for

INDIVIDUALITY

contracting a cold, the flu, or some other disease. Some people refuse to use airplanes, trains, and buses because of the perceived higher risk of contact with other travelers who are total strangers.

Individuals often want or need to drive by themselves, because they have personal or business appointments that require going someplace. The perception of the importance of travel independence often precludes the efficiency of riding with someone in a single vehicle or taking a bus with other persons. Sometimes people have important appointments and they feel justified in making the decision to travel alone. The choice to drive alone sometimes results from actually needing this independence. In the future, this human nature phenomenon will cause a resistance to the use of any current mass transit facility, such as buses and trains.

The HEVER system will exist as a truly high-speed, high volume mass tranport system, emphasizing non-stop operation, greater safety, and higher efficiency than that for trains. Today, passenger trains mass transport relatively large groups of people. In order to compete economically, in most train transport, a mixture of passengers, products, and material must travel on the same train. With the new system, handling a massive number of vehicles, each existing as single units can more easily meet the demanding requirements of reasonable cost for reasonably fast service.

In some cases, buses can actually provide a slightly better alternative to trains. Buses offer a viable option on the future HEVER system. Normally, though, most individuals will want to travel without others invading their private space on a bus or train and without the delay caused by waiting for others. When one needs to get someplace on a non-routine basis, people generally prefer to travel in their personal vehicle. A personal vehicle can take someone to virtually any place at any time relatively quickly, and one can take more of his/her personal items than when using current public, mass transportation facilities. Newly designed vehicles will need to be able to use roadways at a reasonable fuel cost. Low cost (preferably free) parking must be highly available.

Probably most of the time that current mass transit works, the source and destination attract large numbers of people who want to attend the event. Examples of this include entertainment extravaganzas, and popular collegiate and professional sports events. In these cases, and in addition to people driving in their own automobiles, light rail vehicles, and buses transport large groups of people between several pooling points and the entertainment destination or sporting event. When buses or light rail vehicles can make stops at more places where people can enter and leave, they will generally operate more economically. This same phenomenon of needing to stop, significantly contributes to making trips much less desirable; obviously non-stop travel is preferred.

Driving fast, so much fun in my four wheel drive SUV,
I hope you will agree.
Why worry if gas costs so much, I never really cared about mpg.

Chapter 4:
Miles per Gallon

Above 55 mph or 60 mph, currently, a vehicle can either get better miles per gallon or higher speed; it's a tradeoff. If the automobile maximum allowable speed on highways dropped from 70 mph to 55 mph, most cars and small pickup trucks would operate more efficiently in their use of gasoline. The drop in gasoline usage would also actually reduce pollution for the miles driven. Most drivers of vehicles would strongly object because they would feel the pain of trips taking longer, even though everyone's gas cost would be reduced. If federal officials reduced the Interstate legal speed limit to 50 mph, law enforcement officers would usually look the other way as long as driver's vehicle speed remained about or below 60 mph.

Informal (suggested) reduction of speed limits might also accomplish improved gas efficiency by promoting a national informal goal through heavy advertising. This would not require changing speed limit signs. However, probably most persons would not adhere to informal calls for speed reduction and often would even speed above any legal limits. Yellow speed limit signs are basically suggested speed limit signs that indicate danger. These signs recommend that drivers operate their vehicles at or below a particular speed. Drivers regularly ignore the advice on these sign and drive at whatever speed suits their driving style.

Many drivers just want to arrive at their destinations as quickly as possible; they just don't care about miles per gallon (mpg) being high. A few automobile drivers, who want to save money or pollute less, probably would go considerably faster than they typically do except for the perceived constraint caused by the tradeoff of fuel usage and optimal speed to obtain the best mpg usage.

Reducing the maximum speed limit to 55 mph or 60 mph, or lower, for automobiles will generally yield improved miles per gallon. As speed increases over 60 mph, gasoline efficiency gets disproportionately worse. However, speed limit reduction caused by a need for reduction of gasoline usage can sometimes yield a counterproductive mph, especially for eighteen wheeler trucks if they have to down shift to less efficient gears.

Increased traffic volume causes increasing demand for oil based fuel. In the long run, this usually causes gas prices to rise, which negatively impacts our economy and causes more damage to our environment. We should conserve oil for various productive uses other than burning, such as for plastics and lubrication.

According to a Material Safety Data Sheet (MSDS) presented on the Internet, ethanol, a combustible fuel, somewhat toxic in its form before burning, contains 70 percent ethyl alcohol. The MSDS gives ethanol the name ethyl alcohol. It will cause severe problems if splashed into the eyes. Biodegradable in its form prior to burning, the MSDS states that ethanol contains no hazardous air pollutants or Class 1 or Class 2 Ozone depletors.[13]

Ethanol might not directly reduce vehicle mpg, but it would effectively reduce the mpg produced from burning fossil fuels. Burning ethanol fuel helps to save some oil from burning, while reducing pollution, one of the country's important long-range goals. There are issues with ethanol. Currently, most of our nation's gasoline already contains fairly close to a 10 percent amount of alcohol. The ethanol industry does not exist as an entrenched institution; you can't always find gas stations that provide ethanol. Many of the vehicles on the road have engines that will not easily burn ethanol if the alcohol content increases to 15 percent; it will put far too much wear on their engines.

Also there is much controversy relating to impacts that would occur if we allowed the ethanol industry to expand. Very

---

13   Material Data Sheet (MSDS), http://www.nafaa.org/ethanol.pdf.

many replaceable resources, such as corn, weeds, and some garbage provide the materials used to produce ethanol. Various factions have organized on both sides of the issue of expansion of the ethanol industry.

An article, "Ethanol is Facing Key Decision Today," from December 1, 2009, the EPA said it would decide if it will allow raising the limit of corn-based fuel to be raised to 15 percent in gasoline.

Growth Energy, a trade association representing more than fifty ethanol producers, has asked the EPA to raise the limit. The limit has stayed at the 10 percent level since 1978. After the EPA decides, some interested parties likely will end up quite disappointed. Relatively large groups opposing raising the limit state that, at this time, it would have severe negative impacts. Examples cited include automotive and engine manufacturers, food and livestock industries, and oil companies. This article also cited environmental advocates as not wanting the limit raised.

E85, a purer ethanol fuel requires Flex Fuel vehicles; demand for E85 has grown very slowly. At the time of this controversy, the ethanol industry producers had 202 plants in the United States. Growth Energy claims that the capacity of its plants has already reached close to a 10 percent content of ethanol in our nation's gasoline volume and that the ethanol industry cannot easily expand without raising the limit to 15 percent. Also the group claims that raising the limit would encourage increased investment in the ethanol industry, create increased employment, and reduce our dependence on foreign oil. As an interim solution, should the EPA not approve raising the limit to 15 percent, Growth Energy will seek a 12 or 13 percent limit.

The American Petroleum Institute has argued that no credible scientific support exists for allowing more ethanol in gasoline. Also, opponents argue that great damage would happen to millions of lawnmower motors and other similar small machine motors. Grocers indicate that using more corn in producing ethanol would cause higher corn, meat, and dairy prices. According to opponents, most of the demand increase will occur in the Corn Belt states where a relatively low population density exists. On the highly populated east and west coast states, demand will likely grow quite slowly.[14]

Another subsequent article indicated that the EPA is delaying a decision until the middle of 2010, but was leaning toward approval at the time, stating that automobiles manufactured in 2001 and later should easily handle the 15 percent ethanol content in gasoline. The EPA calls the fuel mix E15. Automakers and engine manufacturers praised the delay, indicating quick approval would likely have negative impacts. Delaying the decision probably will give more time to study impacts. However, it is interesting that delaying a decision that would favor ethanol producers will likely negatively impact any farming plans that would have resulted in increased corn production. Basically, due to the likelihood of not enough corn production, if the EPA should allow ethanol production expansion in mid-2010 to happen, expansion of production won't likely happen during 2010. If ethanol producers expanded production, they would have to face the risk that the EPA might not approve increasing the ethanol content limit.

Increasing the limit of the amount of ethanol allowable in gasoline, at best, would represent a temporary solution. It will happen in a few years that gasoline usage will greatly reduce when the electric automobile becomes an institution, the travel mode of choice.[15]

It appears that ethanol won't significantly impact the mpg allotted to gasoline; at least it won't likely happen in 2010. Even if approved, this is a temporary one. The HEVER system will obviate the need for ethanol.

14　Jeffrey Tomich, "Ethanol is Facing Key Decision Today," *St. Louis Post-Dispatch*, December 1, 2009.
15　David Shepardson, "EPA Delays Decision on Boosting Ethanol Blends," *Detroit News Washington Bureau*, date????.

*As I come to this crossing, why should I have
to wait for a darned train?
I will race to beat this train that might waste
my gas, please say it's not insane.*

Chapter 5:
Trains, Trucks, and Airplanes

Locomotives have been around since the early 1800s; horses, livestock, and people originally greatly feared them. During the infancy of railroad locomotives, when they traveled through towns and cities, municipal ordinances normally required that locomotives reduce their speed. In the United Kingdom, the Department of the Environment (DOE) passed the 1861 Locomotives Act which restricted locomotive speeds to 10 mph in towns. The DOE next passed the Locomotives Act of 1865 reducing speeds to 4 mph on open roads and 2 mph in towns, and locomotives had to have a crew of three. The law required that one of them holding a red flag had to walk twenty yards ahead of the locomotive. In 1878, an amendment did away with the red flag requirement, but still required a man to walk twenty yards in front of locomotives. Finally, in 1930, The DOE passed the Road Traffic Act, which repealed the Locomotives Act of 1865 and raised the speed limit to 20 mph.[16]

Needless to say, people in the mid 1800s to about 1930 still greatly feared locomotives somewhat because they were large and noisy. They were new and often perceived as dangerous; people often misunderstood any explanations as to their value versus risk to people and horses. Before railroads, canals and roads (mostly dirt) were the principal mode of transport.

From about 1850 to 1870, congress granted a huge amount of public lands to twenty-six railroad companies. Opponents to the land grants fought the railroad companies in court claiming that

---

16   UK Department of the Environment, "A summary of important legislation," http://www.roadsafetyni.gov.uk/index/road_safety_education/teacherzone-home/teacherzone-mvrus/mvrus-legislation.htm.

the railroads took the land for private company use. They lost; the court ruled against them and based its decision on the "public nature" of the railroad business. Railroads prevailed since the courts ruled that railroads operated as public highways for the people.

Due to an almost feverish enthusiasm for good transportation, many states granted railroads monopoly rights, banking privileges, and tax exemptions. Federal, state, and local governments gave railroads monumental land grants and monetary subsidies.

An excerpt from "Economics of Transportation," by D. Philip Locklin, shows that the government gave to the railroads approximately one-seventeenth of all of the land in the US.

The total acreage patented to railroads was over 130 million acres. This represents an expanse of land equal in size to Michigan, Wisconsin, Illinois, Indiana, and half of Ohio.[17]

While the granting of such a large value caused much of the growth and power of railroads, it was deemed necessary primarily because of the fact that the nation needed (wanted) a link between the East and West coasts. The oil and steel industries would become primary beneficiaries of the explosive growth that followed. Much of the land granted was, at the time, not productive land for farming or ranching. Some of the land was granted for railroads to sell in order to finance the construction of the railroads. Originally hard to sell, land values eventually increased significantly, giving railroad companies tremendous value. Also, the potential impact of the gasoline combustion engine couldn't be imagined at the time that the steam engine became popular. Basically, railroads were faster, glamorous, and could carry larger loads than any other land-based vehicles.

Trains gave significantly faster results than those obtained from use of the far more economical canals. Without railroads, canals would have required more warehouse storage, especially in northern regions, since most vegetation would not survive the cold weather during winters. Also, many canals would freeze in the winter. Materials would take a relatively long time to arrive by

---

17   Locklin, D, Philip, Phd, Economics of Transportation, (Homewood, IL.: Richard D Irwin, Inc.,1972), 134.

TRAINS, TRUCKS, AND AIRPLANES

wagon. Trains could deliver faster, travel long distances, and they could operate during winters. When railroads entered the scene, they basically caused the curtailing of canal development.

Operating more efficiently than wagons, trains found increasing usage. Still, the railroad industry developed in a somewhat complementary way to wagon transportation; wagons would still find use in transporting people and freight to and from train stations. At the time, no other over-the-road commercial transport could compete with railroads for bulk shipment.. This provided some advantages for personal travel by combining freight transport with personal travel service. Even with higher speed, a lower risk of damage or harm was evident. Over the long term, this perceived greater safety, comfort, and speed caused people to eventually understand the improved personal and commercial transportation. Favorable perceptions of these benefits helped to cause acceptance of railroads.

While trains operate, they have the constraint of loading and delivering freight and passengers to and from a limited number of locations. Using sets of tracks that allowed trains to travel so fast also limited their flexibility. While trains are constrained and sometimes relatively slow, they may normally exist as the safest way to travel. Trains have a far less probability to have accidents than automobiles, trucks, and buses. However when a train accident happens, it usually results in disaster.

In an article from the *St. Louis Post Dispatch,* the use of rails allowed a new French-designed train, operating in Korea, to make trips at a much faster speed than the train it replaced. Today, other trains can travel considerably faster. In the Korean case, the increased speed of the train, compared to speed of the train it replaced, caused a reduction of the time for each trip by about one hour and twenty minutes. Assume the new train carries an average of four hundred persons per trip and makes ten trips per day. The average number of persons transported per day would total about four thousand. Multiplying four thousand persons per day times 365 days equals 1,460,000 travelers per year. Multiply 1,460,000 travelers by 1.333 hours saved per traveler equaling 1,946,180 hours per year reduction in travel time for travelers. Using other

31

estimated figures would change the above result, but this makes a significant point in travel time savings.[18]

As railroads basically obviated the need for canals, so did the automobile and truck eliminate much of the passenger and freight transportation demand previously enjoyed by the railroad industry. Truck and automobile industries radically reduced the need for rail transport of people and material goods. Trucks can go virtually anywhere at any time, even though they sometimes are a more costly way of transporting than if one ships by rail. If the product can be shipped by rail and if the length of time to deliver does not raise an issue, transporting by train can offer a lower price, making it the better alternative.

Both truck and rail industries are profit motivated. The rail companies own the rails, locomotives freight cars, and passenger cars. Therefore, railroad companies, having the responsibility to maintain the rail systems and due to bottom line desire for profit, often decide to avoid needed track and roadbed maintenance and upgrades. This sometimes leads to the rail systems getting into poor condition. On more than one occasion, lack of maintenance and repair has caused accidents.

The automobile and trucking industries enjoy an element of low transport cost since they each share their roads and highways and pay only a relatively small amount to maintain them. The property taxes and licensing payments do not amount to a significant portion of total upgrade and maintenance costs. Federal, state, and county government funds for approved projects are supported through collection of traffic fines, sales taxes, personal property taxes, and licensing fees. These pay for most of the highway building and maintenance costs. As the automobile and truck industries rapidly grew, technological advances sometimes improved safety. However, sometimes due to increased traffic volume, with faster speeds, the risk of damage or harm increased. Even so, the perceived benefits of using trucks more than rails, far outweighed the perceived risk.

Many people marvel at the phenomenon of traveling 40, 50, or even 100 mph when driving trucks and automobiles. One can

---

[18] "Bullet Train Launches High-speed Rail Service," *St. Louis Post-Dispatch*, April 4, 2004.

understand that this relatively high speed can present a considerable accident risk, when considering that the automobile or truck travels at a somewhat unconstrained speed and does not have the constraint of tracks. This highly unconstrained phenomenon normally makes the increasing speed of these vehicles more dangerous than transporting by train. Add to this, size and weight, and inertial properties of large automobiles, SUVs, buses, and trucks; this type of traffic can make some situations dangerous, especially for smaller automobiles and motorcycles.

The ability to avoid causing or having accidents requires alertness and a good driving ability, even more so when attaining higher speeds. Even good drivers often cannot avoid accidents and may become unwary victims in accidents. Sometimes drivers do not cause the accidents that they experience; the accident just happens. If a driving mistake, bad road conditions, or vehicle failures, create a higher probability for having an accident, even good drivers will sometimes have an accident.

Professional racecar drivers are highly trained and highly experienced. Racing vehicles normally operate at speeds in excess of 200 mph. Race car drivers, who are usually good drivers, sometimes have accidents on and off of the racing track.

On December 17, 1903, Orville (the pilot) and Wilbur Wright accomplished the first real successful airplane flight when they first flew their biplane for twelve seconds. Through the years, airplanes have meant a lot to our country. Prior to the outbreak of World War I, the government already encouraged mass production of the airplane. World War I broke out in 1914 causing a dramatic increase in the advancement of airplane technology and production volume. By the 1930s, millions of passengers used airplanes.

World War II also greatly affected airplane technology advancement and the massive numbers produced. The first two atomic bombs were dropped from airplanes causing the end of the war with Japan. Today, the NASA Space Shuttle hurtles into space, basically as freight, hitching a ride on a rocket. After completing its mission, the Shuttle returns to earth as an airplane.[19]

19   Heather Whipps, "How the Wright Brothers Changed the World," August 11, 2008, http://www.livescience.com/history/080811-hs-wright-brothers.html

Generally, airplanes transport a relatively large number of persons and often for a greater number of miles per trip than most other modes of transport, and they have a significantly high safety record. Airplane companies generally have to meet high standards for plane maintenance, inspection, and monitoring of flights. Also highly trained airline pilots generally fly far better than most automobile and truck operators drive.

Airplanes took some freight business away from the trucking industry and some freight and passenger business away from the rail industry. Air travel gained its place as an institution because airplanes typically operated at a much faster speed than land based vehicles. This higher speed resulted in a positive impact for trips of longer distances. Air travel costs relatively more than that for either trucks or trains for similar distances, even more so when flights are non-stop. Currently, the high speed employed to transport people justifies the use of airplanes. Also, high value, light weight, low volume freight payloads justifies the use of airplanes. Freight transported is usually made up of relatively small packages. US based commercial jet airplanes normally fly about 300 mph to 600 mph. The unconstrained (somewhat uncontrolled) phenomenon of large, heavy, hurtling objects such as airplanes, automobiles, and trucks yields sometimes high probabilities for accident occurrences for travelers.

If you fly a private, propeller driven plane with good aerodynamics, higher speed, to some degree, can give you an element of greater safety. If your engine fails or you run out of gas, the higher and faster you fly, the more time and distance you have to find a good landing place. If your airplane is going fast enough, you can actually increase your airplane's height above the ground even though your engine has quit. Except for controlled airspaces, airplanes operate virtually anywhere in the air that the pilot decides to fly.

If the number of airplanes flying totaled anything like that of vehicles on roads, the accident rate would give horrible results, far exceeding that for automobiles and trucks. The relatively low number of airplanes along with high airline company standards and federal standards helps to keep the number of accidents low. However, when an airplane accident occurs, it usually ends in disaster.

Some say our Interstate isn't so old;
perhaps to private industry, it should be sold.

Chapter 6:
Our Fifty-Plus-Year-Old Interstate System

Accesses to our Interstate system, the number of lanes, and types of interchanges have developed in a patchwork manner over the past fifty or so years. In highly populated areas, we use a relatively significant amount of real estate for road space Monday through Friday, for a few hours each day. We refer to these periods of time as rush hours. In many cities, a lunchtime rush hour also happens almost every working day. In addition, sporting events or other entertainment events can cause traffic jams due to everyone wanting to go to the events at virtually the same time. For security reasons, a similar traffic problem occurs when a dignitary such as the pope or our country's president or vice president visits a city. This causes unusual traffic problems to occur because of the necessity for the police to close roads temporarily for security purposes.

The relatively low use of our system of roads during non-peak times creates a situation in which we actually use a small percentage of the available roadway space. This phenomenon causes the lowest roadway use at night since most people sleep at night.

Although, often the public or other authorities can approve addition of a lane for a short distance somewhat easily, it usually doesn't help very much. Adding lanes at one Interstate location often exacerbates the overall situation by creating problems where the roadway needs but doesn't get correspondingly widened. As changes to the network of Interstate roads adds lanes, adds or upgrades entrances, or requires other changes, the complexity increases, making it harder to improve roadways in the future. This inefficient roadwork usually still provides the same relatively slow average speed during high volume periods of Interstate usage.

The Interstate system of highways, one of the most significant advances in the history of transportation, still exists as our most highly used and most useful transportation system. The Interstate system, approximately fifty years old, in many situations has overloaded its lanes (for example rush hour traffic) and growth makes it increasingly prone to accidents. Expanded, redone, and patched to the point of being a large, aged system, Interstates have grown costly and difficult to maintain. Difficulty making repairs or improvements constricts flow of traffic especially during construction. This causes relatively high motorist costs while these changes are done.

High traffic volume entering an expressway often causes fear, aggressiveness and a relatively high probability for the occurrence of accidents. When a high number of vehicles enter an already heavily traveled expressway, traffic inherently slows overall, sometimes to a crawl or even stops. This tends to encourage increased jockeying for position and subsequently increases the number of starts, slowdowns, and stops. Ultimately, this must increase frustration and irritation. The sudden aggressiveness causes an increase in the probability for the occurrence of accidents. Drivers often become increasingly impatient and aggressive as the cutting off other vehicles occurs. This phenomenon occurs with no apparent fairness. This can create a chaotic situation and often occurs daily, and often at the similar times and places. Implementation of a computer controlled system would totally eliminate the possibility of the occurrence of this type of excessive competition.

An article from the *St. Louis Post-Dispatch* by Runge, a group that studies possibilities for reduction of vehicle accidents, illustrates that many people are killed annually. Runge seems to imply that society should prevent these accidents. During 2003, 42,815 people were killed in the United States, in road accidents. Runge is part of a U.N. coalition that is trying to raise awareness of human factors that kill people while driving, such as speeding, drinking alcohol, and neglecting to wear seat belts. A separate UN group works to standardize vehicle safety features. "They're not random acts of God," Runge indicated about traffic accidents. "They're predictable, and therefore they're preventable."[20]

---

20   "In the United States 42,815 People Were Killed in Road Accidents Last Year," *St. Louis Post-Dispatch*, Monday, April 5, 2004.

Who causes these great large obstructive blocks?
They make me so late. Don't people ever look at clocks?

Chapter 7:
Great Large Obstructive Block (GLOB)

A Great Large Obstructive Block (GLOB), a concentration of vehicles, sometimes due to accidents, construction, rush hour ,or one or more relatively slow moving vehicle(s), usually causes traffic slowdowns and stoppages. A GLOB of vehicles, sometimes followed by a gap of no vehicles, slows traffic and causes inefficient road usage. As a general rule, humans cause GLOBs.

Relatively slow moving vehicles often create increased probability for other vehicles to become part of a GLOB. This causes abnormal delays and increased probability for accidents. One vehicle on a two-lane road or two vehicles on a four lane road, highway, or Interstate roadway traveling at the same speed can cause the creation of a GLOB of vehicles if the slow vehicle travels below or even at the speed limit. Other faster operating vehicles find it difficult to pass them and they tend to bunch up behind them. This results in relatively low usage of the road space in front of the GLOB. Vehicles in and approaching the GLOB must often travel at a low inefficient speed.

A similar phenomenon exists when a truck transports a wide load such as a mobile home. While it usually does not use more than one lane, often a vehicle transporting a wide load can occupy a single lane so fully that it appears to travel out of its lane. This makes most drivers afraid and slow to pass. Since wide load vehicles are so wide, they often have to use part of an additional lane, which then can cause GLOBs of vehicles. Slow moving buses, trucks, and recreational vehicles can slow down enough to cause GLOBs of vehicles.

Bridges exist as natural bottlenecks; they often create GLOBs. Relative to slowing down traffic, bridges can cause problems similar to other situations involving unduly slow moving vehicles. At times of high traffic volume and regardless of their width, bridges often cause traffic to slow down, creating a GLOB of vehicles leading up to and going over the bridges. This situation creates GLOBs probably because of natural fears of heights, limited width of bridges, and the desire of some travelers to see the sights below the bridge, as their vehicles pass over, for example, a river. When the river below has flooded, drivers of vehicles will slow down even more. Bridges should have at least one extra lane on the bridge and an extra lane for some distance after the bridge to help compensate for the tendency of bridges to cause bottlenecks. Large vehicles such as trucks, buses, motor homes, and other wide loads exacerbate bridge bottleneck situations. Using a computer controlled transport system would eliminate the possibility of the occurrence of this type of delay and gaper blockage caused by fear and desire for entertainment of the situation.

During high volume traffic situations, vehicles' average speed on expressways would not likely vary much by increasing or decreasing speed limits. Expressway construction projects, accidents, and gaper blocks highly impact average speed in a negative manner. This can cause GLOBs of vehicles. No matter how many lanes exist or whatever the legal speed limit, GLOBs of vehicles on expressways will occur all too much. Each addition of a lane yields a diminishing return of efficient usage when compared to that for other previously constructed lanes. When construction projects add a lane, increasing the roadway width, pockets of unused roadway area will occur. Due to increasing roadway complexity caused by the widening of roadways, when traffic volume increases, the blockages may even get worse.

Average drive time will likely decrease in situations of low traffic volume or non-peak drive times. However, overall average driving time will not necessarily decrease in proportion to the raising of a speed limit. In addition to normal heavy traffic, there are many situations such as sporting events, popular entertainment events, and political rallies that will cause additional heavy traffic. Human beings normally do not have enough time to react to fast-

GREAT LARGE OBSTRUCTIVE BLOCK (GLOB)

er and more complex situations and it seems that more accidents and more severe accidents will occur.

With higher speeds, drivers will not likely increase the distance between vehicles in proportion to the increased speed. Drivers still won't want drivers of other vehicles unfairly getting ahead of their vehicle. If drivers maintain too short of a distance between them and the vehicle in front of them, vehicles needing to enter the expressway will often experience delays. Drivers of vehicles already on the expressway will sometimes need to lower their speed or move to another lane, in order to create enough space for vehicles to enter the expressway safely. In some situations today, because of the high volume and high speed on an expressway, traffic signals exist to cause a short delay of vehicles entering the expressway. These signals usually alternate between red and green in order to regulate each vehicle needing to enter the expressway. This allows vehicles at expressway entrances to enter the expressway one at a time with a delay between vehicle entries caused by the traffic signal lights.

Even with higher speed limits, GLOBs will still occur. Ironically, the GLOBs will actually have a greater negative impact on average speed if the roadways on which vehicles travel have higher speed limits than when they have lower speed limits. After getting past the GLOB area, some drivers will go over any speed limit in an effort to make up for the lost time. During high volume times, average speed will likely decrease, sometimes significantly. High traffic volume, construction, vehicle breakdowns, accidents, and gaper blocks will add increasing danger when speed limits are high. While adding more lanes usually, to some degree, will help to improve traffic flow, due to increasing the complexity of the roadway, its effect can be marginal in its impact on increasing average speed or reducing risk of accidents. After adding a lane(s), and because of the ability for drivers to cross more lanes irresponsibly, accident risk is still great (sometimes greater).

On expressways with only two lanes in a given direction, if two eighteen wheeler trucks pull along side of each other and they travel together, side by side, for long distances, they often will cause a GLOB of vehicles. By moving in parallel, trucks often cause the GLOB of vehicles to occur because most other traffic tends to move at a faster speed than trucks. Operation of trucks, while they travel on long uphill distances, exacerbates the situation even if they are not traveling side by side. The result is an eventual traffic slowdown that often takes considerable amounts of time to clear out even after the trucks cease to cause the bottleneck directly. The drivers of vehicles delayed by this GLOB of vehicles often become somewhat angry or at least frustrated because these delays seem unnecessary.

Many trips take longer because of these delays, and it can make traveling less enjoyable. But worst of all, the trip can be more dangerous. The same problem occurs if three lanes are occupied by three trucks traveling in parallel. Sometimes significant numbers of trucks travel in groups (convoys) causing traffic problems by different trucks passing each other. This can create dangerous situations depending on whether the trucks pass while going up or down hill or if they carry heavy or light loads. Sometimes a previously passed truck picks up speed and passes the truck that passed previously. Sometimes it appears that truck drivers treat passing each other as a game.

When an eighteen wheeler travels in the passing (left) lane for long distances, a dangerous and inefficient traveling situation happens. When going uphill, a truck usually cannot go as fast as on a flat or downhill run, especially when heavily loaded. This often causes other drivers of vehicles to wait behind a slow moving truck; sometimes drivers will use poor judgment and even pass to the right of the truck. During this abnormal driving action, automobiles or other trucks cannot be sure that the truck won't move back into the right lane as one gets ready to pass unsafely on the right (bad choice).

Drivers of automobiles or other trucks are almost as unsafe when they try to pass trucks by using a passing lane or take too much time while slowly passing for long distances and for no good reason. However, the large size of eighteen wheelers can create difficult situations where drivers following them cannot easily see around them. Some drivers, mostly automobile drivers, drive long distances in the passing lane due entirely to habit or preference. This type of GLOB creates increased danger and the inconsiderate action can cause anger and frustration.

Stop, stop, why must I stop for all those signs?
Oh, I know; I don't want any more fines.

Chapter 8:
Information Sign, Stop Sign, Traffic Signal

Although not always the case, highways and city streets should have highly visible signs with clear messages, easy to read and understand. The position of these signs should give enough time and visibility for drivers to react properly and not miss a highway entrance to or exit from an Interstate highway. When drivers erroneously pass up a required highway entrance or exit, the driver's presence of mind can easily shift away from concentration on his driving. He must then shift some of his thinking onto the problem of his now erroneous location caused by his error. He must then determine how to get back to the correct direction.

This usually results in the wasted time caused by the error. The magnitude of time lost roughly equates to the amount of time expended until the error is discovered plus the similar amount of time that the driver must waste essentially undoing the error. Concern for safety can easily decrease, giving way to the driver paying more attention to the error just made. He will then need to find the proper roadway and return to driving in the correct direction toward the intended destination. Making direction change errors wastes time, causes confusion and irritation, and sometimes can cause creation of a dangerous situation. We should, and probably will, make roadway direction change information more automated in the future.

The presence of stop signs and traffic lights commonly occurs in towns, cities, and at some rural locations. One can find intersections that have no stop signs or stop lights where low enough traffic obviates their need. Occurrence of increasing high traffic volume at an intersection usually causes a perception that a greater probability for accidents exists at the intersection. Speed, number of accidents, and increasing volume will then often lead to installation of a stop sign (non-lighted) to help slow traffic down and theoretically reduce the probability for occurrence of accidents at the intersection.

Speed, number of accidents, and increasing traffic volume can cause the perception of an increased accident risk, thereby causing installation of a traffic light to occur. Further concern for increasing accident probability can cause the widening of streets and/or lowering of speed limits to happen. Increase in traffic volume causes growth of more complicated roads and intersections and results in an increased probability that accidents will occur. This type of problem usually doesn't go away completely by installing more stop signs or stop lights, widening roads, or doing combinations thereof.

The existence of multiple stop lights in a short distance often causes congestion, leading to competitive aggressiveness and frustration due to the delays. This situation has a relatively high risk of accident occurrence. Many Interstate overpasses with their accesses to and exits from the Interstate built twenty to fifty years ago, have consisted of design and construction with consideration of relatively low traffic volumes that existed at the time they were built. Due to old design, these overpasses, no longer adequate, have now even become accident prone. Designers have employed engineering methods of past times. Even at the original time of design some Interstate overpasses and accesses were inadequate. Many of those Interstate intersections whether constructed recently or years ago, currently do not allow good traffic flow. Inherently they exist as bottlenecks.

City officials sell blighted public owned land for two dollars.
For $2.3 million they need to buy it back, do you hear any hollers?

Chapter 9:
Interstate Design and Implementation

Occasionally an Interstate overpass (non-cloverleaf) has entrances and exits in both directions and a side road on each side of the Interstate, both side roads paralleling the Interstate. Overpass builders currently ordered to construct this type of overpass, built pretty much like those in the past, place the overpasses in situations where they usually experience relatively low volume. Government officials should champion better original design when building new overpasses. The Interstate system also needs upgrades of many of our existing overpass intersections, especially the older ones where increased volume has built up through the years. However, the upgrading of overpasses will be less needed if the country builds an elevated computer controlled roadway.

Often when an opportunity to get approval for a needed upgrade arises, politicians and voters do not approve the change, causing continuation of a less than desirable traffic situation. Much of the political and voter resistance causes rejection of upgrade proposals due to the perceived high cost to satisfy today's perceived upgrade importance. Ultimately, society should upgrade some of these overpasses. This is because, as the conditions worsen with increased traffic flow, the cost to society increases in the form of increasing the burning of gas, wasted time due to increasing traffic delays, and more accidents. More severe traffic conditions often cause delays to increase due to restricted flow during construction periods and while the upgrading of these intersections actually takes place. Also less traffic efficiency during construction periods causes a worse driving situation due to existing high traffic volume that caused the need for the upgrade in the first place.

The high cost of obtaining additional land to improve the traffic situation can deter gaining approval for many needed upgrades. Sometimes, strong resistance from landowners makes it virtually impossible to accomplish the needed change. Eliminating this long-term time bomb by originally doing a more farsighted design and implementation of some of the overpasses would have resulted in lower gas usage plus a safer and more enjoyable roadway.

Admittedly, this is twenty-twenty hindsight. Authorities often find it harder to sell the public on a better, usually bigger, overpass because of the obvious lower cost of the smaller overpass option. Originally designing and building these overpasses with more consideration given to future costs and traffic difficulties, the future would have resulted in a better long-term traffic situation during the following years of usage. Cost of ownership, including delays, accidents, and overpass replacement, along with other upgrades, increased land costs, and maintenance through the many, many years can be quite high. Higher ownership cost often occurs in the long run. This cost negatively impacts stakeholders considerably more. The stakeholders include drivers, but, from a cost standpoint, they consist mostly of the tax-paying public. Finally, a far better overall driving experience would have resulted if better designs were initially implemented.

Assume an Interstate overpass situation where outer roads parallel the Interstate on both sides of the Interstate. Further, assume that entrances and exits (non-cloverleaf) exist in both Interstate directions and on both sides of the Interstate. In this situation, the interstate overpass intersection will have as many as four intersections with sets of associated traffic lights in a few one hundred feet. The traffic situation requires turns and associated traffic lights at each of the four intersections with each intersection usually requiring left and right turns. A nightmare often occurs during high volume traffic. No one apparently perceived future high traffic volume as an issue when originally designing and building the Interstate overpass.

INTERSTATE DESIGN AND IMPLEMENTATION

Many Interstate overpass intersections similar to the above non-cloverleaf overpass do exist. Such a set of Interstate intersections actually does exist in O'Fallon Missouri. Instead of constructing an overpass with entrances and exits connecting directly to a cross road, now the busiest road in the city, the Interstate intersection could have had entrances and exits connecting with the outer roads, instead of directly to the busy cross road. This would have reduced the number of intersections to two instead of four and the number of places at the intersections requiring turns to one half of the current number. This would have reduced electricity usage and cut in half the number of required traffic lights. This situation has existed for well over twenty-five years. Had developers originally built the intersection properly or even upgraded it years ago, the people who drive on this busy city road through the city of O'Fallon would have experienced lower cost and improved traffic on the city's busiest street.

The above Interstate intersection normally has high traffic volume. This often causes an extremely congested situation, sometimes even under relatively low traffic volumes. Due to normal high traffic volume, the city has set relatively low speed limits at this complicated set of intersections. Traffic on the road going under the Interstate usually moves quite slowly, intersections sometimes gridlock, often drivers jockey for lanes, and tempers flare. This continues, often creating situations where there remain relatively high probabilities for the occurrence of accidents to happen.

An article, "$2 Sale of Land Proves Costly," from the *St. Louis Post-Dispatch* demonstrates how the difficulty of managing vacant, blighted land can cause its sale by government officials at a ridiculously low price. The city of St. Louis sold the land in question for $2.00, that's two dollars, and as of the date of the article must pay $2.3 million dollars in order to get the land back. The bridge project, now fully approved, cannot happen until this land is obtained by the city of St Louis. At the time of the sale by the city, some officials apparently thought that by selling the land, new owners would improve it and restore the blighted area, therefore justifying its sale.

Other people seemed to question the validity of the sale because they believed that the project for building the bridge would, without question, happen in due time. One can see the difficulty of somewhat large, especially government, organizations, managing property and also the difficulty these organizations have when they must secure usually scarce property for construction before projects can get approved and developed.[21]

---

21   Heather Ratcliffe, "$2 Sale of Land Proves Costly," St. Louis Post-Dispatch, Monday, 10/18/2009.

I love change. Change is great, don't you agree?
Please, please, please, don't change me.

Chapter 10:
The Interstate Champion

During the immediate years after World War II, our country exhibited a happy, almost exhilarating, positive mood. During the mid-1950s, in spite of dealing with a Democratic party that controlled Congress, President Eisenhower championed and gained congressional approval for building the Interstate. Virtually everyone knew that the Interstate implementation would take a great amount of time and cost to implement. The country bought into the idea of creating the Interstate even though we already had the best transportation systems in the world. The people accepted it as the right thing to do in spite of the Interstate's obvious high cost to design and build.

If we had not built the Interstate in the mid 1950s and 1960s we probably would not have grown strong economically as soon as we did. At that time, if we had procured (then) cheaper and more available land, we could have built a far greater number of overpasses with cloverleaves and outer roads. Minimally this would have allowed later upgrading from non-cloverleaf to cloverleaf intersections and they would have happened at lower costs. Due to not procuring more land in the 1950s, building almost anything of an expansion nature now proves far more difficult and costly.

Interstates have their share of congestion. In general, the original design gave virtually no consideration for the need for extensibility as a major future requirement. In the mid 1950s, developers knew fairly well that significant population growth would happen by the year 2010. They should have created the Interstate with more foresight. Population growth has caused traffic congestion and, therefore, resulted in the need to add more lanes to the original number of Interstate lanes as traffic flows inside of and toward cities. This same phenomenon occurs for traffic flow leaving cities. The current high volume of traffic flow congestion shows a lack of foresight in the original planning for future traffic as it flows into an Interstate at its intersections. Because of the somewhat non-coordinated upgrading of the Interstate, the happening of new construction has often caused poor traffic flow conditions to continue to occur.

Growth through the years happened mostly as a path of least resistance, now making it virtually impossible to design upgrades that easily integrate with the existing Interstate road systems, especially in large metropolitan areas. Yet our Interstates have historically enjoyed the best overland movement of humans and material in existence.

This happened partly due to the fact that private industry does not own the Interstates. If private industry owned the Interstates, as it owns the railroads, the Interstates would in no way have resulted in as good of an experience. Along with government subsidies, private industry would likely have obtained revenue through the heavy use of toll booths. Maintenance and upgrading would have required balancing cost against revenue collection and profits to owners. Similar to lack of maintenance by railroads in the past, private industry would have sometimes avoided proper maintenance of Interstate roadways. They would likely have done the maintenance even less timely than politicians get it done.

Approving a major (and sometimes minor) Interstate change often requires a monumental amount of political effort. Usually, the smaller the patch, the easier government officials can understand and sell it politically. Voters and government officials normally cannot understand a request for too complex of a change. The political approval process usually represents only a part of the process. Far from trivial, the total process includes understanding how to recognize a valid need for change, design the change, get approval, and properly implement the change. The change might not appear important, feasible, or practical, and, therefore, along the way approval might not happen.

THE INTERSTATE CHAMPION

Travel without the Interstate system cannot compare to after it had initially been designed, built, and implemented. The newly developed Interstate system provided unbelievably more efficient, safe, and enjoyable travel. It handled traffic acceptably, at relatively high speed, and also significantly reduced the amount of time spent traveling. Due mainly to the non-stop nature of the Interstate system, drivers of automobiles and trucks could travel, at a relatively high-speed and arrive more quickly at desired destinations. Virtually, this eliminated all of the stops required on the old highway system. Appeal of the positive nature of Interstates, urban sprawl and population growth created the overwhelming volume of traffic that now travels on the nation's Interstate roadways. The number of lanes originally built was normally two or at the most three. Eventually, after the original implementation of the Interstate system, the need to add lanes occurred. Quickly these additional lanes often become saturated shortly after building them, sometimes as soon as developers built them.

When using current roadways, one can easily recognize chronic bottlenecks; it doesn't take a rocket scientist. However, we often live with these bottlenecks for decades. On city Interstates, during high volume traffic situations, it is relatively easy to know where stretches of roadway exist in which traffic almost always slows to a crawl or stops. In the same kind of high overall traffic situation, like rush hour, one can see that low traffic volume also sometimes exists, causing an inefficient and low use of some stretches of Interstate roadway. Sometimes one can see that almost always there is no traffic at all. This suggests an inherent lack of good flow designs in our expressway systems. This situation, in which builders originally provided an adequate Interstate roadway, met with increasing traffic volume. Bottlenecks eventually occurred more often.

Due to existing structures that prevented developers from obtaining the needed land for roadway change, officials often had to judge the project as infeasible. Persons responsible for approving changes often viewed it as too costly or almost impossible to obtain the needed land from owners. These bottlenecks usually happened and their number grew because of increases in traffic

HEVER

volume and population and sometimes actually due to poor design or poor execution of constructing approved changes. We currently have good, quite usable road systems. However, due to the increasing volume of vehicles and road patches, an overload of our Interstate roadways has occurred, causing the system to often break, so to speak, from its own weight.

To build cloverleaf overpasses, construction cost is such a high price.
Let's not worry about the future, a low price now, that would be nice.

Chapter 11:
Driving the Interstate

When one considers a portion of about thirty miles of the mostly non-elevated Interstate system going through suburbs and the heart of a city, one should easily visualize the virtual impossibility of attempting to build, manage, and provide safe navigation of an Interstate system that will stand the test of time, mostly because of an inherent occurrence of growth. Our country needs to design and build a highly computer controlled, completely elevated roadway transportation system that will complement the current Interstate system. Some very good drivers and some not so good drivers attempt to navigate our current Interstate system. Often these drivers have poor health, poor coordination, and sometimes they drive while under the influence of drugs and alcohol. Often error-prone, relatively slow at reacting, and sometimes actually frightened drivers cope with driving on our complex set of Interstate roads.

The normally fast lanes can sometimes develop into relatively slow lanes. Drivers of vehicles tend to cut over into the leftmost (thought of as the fastest) lane. Quite often, this lane does move the fastest but variation in speeds of vehicles in traffic lanes can cause a considerable amount of competition for getting into that fast lane. Another element, driver variations in aggressiveness helps make some impact and can cause high levels of competition for the fast lane. Congestion in the normally fast lane depends on each driver's driving style. Drivers of vehicles often compete more aggressively for available, perceived as unused, roadway space during times of high traffic volume, such as rush hour. This can result in all lanes experiencing slow down and sometimes stoppage. Perceptions give the impression that one will make the fastest driving time by using the normally fastest lane. This sometimes causes the fast lane to become slower than other lanes. Variations in the speed of vehicles and human nature (poor driving ability and/or lack of attention to driving) can cause problems. In the future, by using a highly automated, computer controlled system we will eliminate the occurrence of this type of excessive competition while having our vehicles transported on such a system.

City and suburb Interstate roadways often experience bottlenecks due to increasing volume. Often the number of lanes will need to increase in many places as the roadway extends itself to and from the rural and suburban areas, through the suburbs, and eventually to or through the city. As the number of lanes increases in one area, often they do not increase appropriately in other adjacent areas. Adding or widening entrances at one location can also increase congestion in other different places on the Interstate system, after developers finish making the actual changes. Currently no one can know all of the impacts caused by changes after construction personnel make the changes to our Interstate road systems.

During rush hours, as traffic flows from the suburbs into the city in the morning and in reverse in the evening, bottlenecks often will occur. The disproportionate number of Interstate lanes and Interstate entrances flowing into a smaller number of Interstate lanes usually causes predictable bottlenecks as they wind their way through the city and the suburbs. Places on Interstates exist where five or eight lanes flow into a fewer number of lanes, even into as few as three or two lanes. Sometimes this will happen in an extremely busy portion of an Interstate.

Also, some situations such as bridges and some poorly designed entrances and exits will exacerbate the situation. Of course, the perfect system cannot in reality happen but it certainly could come closer to perfection if politicians and other persons with special interests did not impede approval of obviously justifiable proposals. As such, they sometimes block needed solutions to traffic congestion problems. Improving the Interstate system often takes tremendous political effort and usually by the time the political arena understands and/or accepts a change (one can think of the change as an Interstate patch), the change does very little to help relieve traffic problems. Even when making changes within a reasonable time, they usually do not do much to reduce traffic problems.

After adding a lane or two, the traffic will still bunch together. There will still be construction, sunshine slowdowns, accidents, new or maintenance construction, automobiles cutting across lanes, and gaper blocks. There will be inconsiderate, careless, reckless, incompetent, and sometimes drugged or drunken drivers. When anything goes wrong on the Interstate system that impedes traffic flow, many vehicles will need to slow down or even come to a long screeching halt. Delay of hundreds of vehicles is not unusual, often for hours. This can cause an inability for vehicles to move or exit the Interstate in order to take an alternate route. Sometimes, when vehicles' drivers can get off of the Interstate, away from the blockage of a slowdown or stoppage on the Interstate, no practical alternate route exists. In this case, the Interstate travelers must wait for the blockage to clear. This costs valuable time and sometimes money and lost opportunities. Even after clearing a blockage, traffic tends to move slowly for a considerable length of time.

Construction inherently causes (sometimes severe) traffic delays. When large construction equipment moves across an Interstate roadway, traffic will need to stop. Construction activities can require the forcing of roads to shut down completely, sometimes for extended periods of time. This will often cause drivers to use unfavorable alternate routes. This in turn causes increased traffic volume and congestion on those alternate routes. When construction repairs occur on one or more lanes of a three or more lane Interstate, often an additional lane is closed for construction logistics and safety purposes.

Execution of construction projects, accidents, stalled vehicles, and slow vehicles are all causes of traffic delays of any vehicles traveling in the same direction as the traffic directly obstructed by such a situation. Often, the above blockage also causes another, secondary traffic delay or stoppage of vehicles that would normally go in the opposite direction. One usually refers to this as a "gaper block." It results in a slow down or stoppage of vehicles as they pass a traffic situation where vehicles or some other item of interest has caused a slowdown or stoppage. This type of problem can happen because of an unwise interest in the situation by some drivers who then slow down enough to feel safe enough, while they try, out of curiosity, to see as much as possible. Usually, one cannot actually see that much while driving a vehicle past the mishap or congestion. Ironically, gaper blocks often cause secondary accidents due to driver attention directed toward the situation causing the gaper block. It basically tempts the driver to "get a good look"; this usually causes traffic congestion to worsen.

Problems of a political nature and other types of hindrances do sometimes prevent the upgrade of Interstates. Some approved proposals actually shouldn't get approved because they lack cost effectiveness. They sometimes get approved for poor reasons, available money or due to personal favoritism. Construction consisting of maintenance, repairs, or upgrade of Interstate roadways sometimes causes a marginal or totally wasteful result. This often causes extremely counterproductive results in the sense that road construction funds could have found use on a better project. Travelers, positively impacted by a better use of these funds, would have experienced more travel benefit.

Sometimes drivers, using poor judgment, continue at high speed even under abnormally high traffic volume. Drivers should maintain a safe distance between their automobile and the one in front of them. The higher the speed of their automobile the greater need for increasing the separation distance. Lane changing or jockeying for position often causes a resulting dangerously short separation distance. In most high volume traffic situations, drivers often do not maintain safe distances when behind vehicles. Often, drivers make an unwise decision, in which he or she does not leave enough separation space in front of his/her vehicle, worrying that another driver will unfairly cut in front of his/her vehicle to gain position in traffic. Drivers do this aggressive cutting into traffic in an effort to get to their destination more quickly.

The danger in these situations increases when a driver of a large vehicle prevents the driver following his vehicle from viewing the roadway in front of his obstructing vehicle. The driver of the following vehicle often does not maintain a safe separation distance. In this case, one cannot see if a safe space exists or doesn't exist in front of the large obstructing vehicle ahead. If drivers of vehicles farther ahead must suddenly slow down or stop, less available time to react causes the probability of accident occurrence to increase. Driving behind large pickup trucks, SUVs, large eighteen wheelers, or vehicles with wide loads such as prefabricated housing creates a far more dangerous situation than when driving behind smaller vehicles. Accidents often happen due to the irritation of following large, sometimes slow, vehicles or by trying to pass them. People often erroneously blame other reasons for the accidents. In high traffic volume situations, probability for accidents increases as vehicle separation distance decreases, speed increases or visibility decreases.

Perceived increased risk of accident occurrence can cause speed limit reduction. The speed limit reduction has a marginal effect since reducing speed increases the length of time vehicles are on the road for any given distance. Reducing the speed limit contributes to congestion in high traffic volume situations and is usually unnecessary in low volume situations. Sometimes reducing speed limits causes creation of speed traps.

It is not unusual for a highly used, somewhat rural two-lane highway to have a speed limit of 55 mph or 60 mph until a three-, four-, or five-lane road replaces it, at which time officials might actually reduce the speed limit to 40 mph or 45 mph. With the improvement of a road's width comes a perception of growth of increased driving hazards due to new crossing roads, new subdivisions, or business growth and ,thus, greater perceived possibility for accidents. Even with the better roadway, this perception of risk will often cause speed limits to drop rather than stay the same or increase.

Some drivers inherently exhibit a need to jockey for position in traffic. In an all too often inclination, many drivers apparently, basically want to compete, even to race. They seem to want to get a better position in traffic, mostly to get ahead of one or more other automobiles. Sometimes drivers get one or two automobiles ahead and then suddenly dart across several lanes to gain a better (perceived) position in traffic. Some of these drivers will race to get ahead of one or more automobiles before immediately exiting off the expressway. Also, sometimes a driver will race dangerously to get ahead of a vehicle entering an expressway, only to then use an exit almost immediately. Often this happens even when no traffic follows the automobile that is entering the expressway. A mind set (mental state) or a way of thinking exists that brings drivers to feel a need to get ahead of just one or two more automobiles.

Would average speed increase with speed limit increases? Obviously the answer is sometimes yes, sometimes no, usually depending on traffic volume. Regardless of speed limits, during peak rush hour driving times, annual average speed on expressways, when considering slowdowns and stoppages, typically can range between 20 mph and 30 mph, more or less. With the addition of needing to drive slower while driving through subdivisions before entering and after leaving expressways, average speed of home-to-work trips will not likely have a speed much greater than 20 mph.

While a great abundance of statistics exists about drivers, speeds, and accidents, these statistics have not lead society to address how to solve the problems caused by making roadway changes or by the traffic complexity growth caused by increasing traffic volume. In general, our expressways have more volume than they can effectively handle and imperfect human beings do the driving. Humans will still accidently or on purpose ignore traffic safety requirements and break traffic laws. We have overloaded our expressways, making them almost impossible to expand and increasingly difficult and expensive to maintain.

Often drivers navigate our overly busy streets with difficulty, for instance when making left turns. Driver attitudes vary relatively as to how quickly he or she decides to enter relatively fast moving traffic while the driver's vehicle starts from a stop or while his or her vehicle moves quite slowly. A slowdown or stoppage usually occurs when drivers use ramps to enter into Interstate expressway traffic and the ramp traffic has slowed enough to cause a backup on the ramp entrance. This often results in the occurrence of a risky situation due to different degrees of caution in drivers when they enter a faster moving roadway from slower moving traffic on an entry to an Interstate roadway. A similar phenomenon to that exists where people must slow down, for example, for a bridge, construction, an accident, or a gaper block. At a minimum, this causes irritation and frustration.

Perception of relatively slow speed within a subdivision can give an impression of high speed, while high speed on an expressway can give an impression of slow speed; perceptions rule us. Vehicles could go much faster than allowed today. However, there exists the perception that speed rather than human action causes accidents. Most of the time drivers do cause or at least have accidents. Some accidents happen accidentally without anyone at fault. There exists a common belief that in the case of an accident, a higher probability of death will occur when people drive at a higher rate of speed. When accidents happen, one of the major elements, driving at higher speeds, does contribute to higher probability of death occurring. While driving on roadways, other factors also contribute to the occurrence and severity of accidents, such as faulty vehicle components, not wearing seat belts, consuming alcohol beverages, using a cell phone, eating, fixing one's hair, and powdering one's face.

An article in 2006 from the *Washington Post* showed a rise in roadway deaths to its highest level in fifteen. A total of approximately 43,000 people died in traffic accidents on US roadways. Further, the article indicated that drivers drive faster making the traffic environment more dangerous. The number of people driving under the influence of alcohol continues with no reduction in its negative impact. To some extent, this exacerbates the situation for motorcyclists and pedestrians. In many states, after the 1970s, several states revoked laws requiring the mandatory wearing of helmets by motorcyclists. In these states, motorcyclist traffic related deaths rose sharply. Statistics at the time showed that more men than women die. Although this information is old, it shows that human beings make decisions that impact the potential for accidental death. Also, as long as people (sometimes artificially impaired by drugs or alcohol) control vehicles and drive at high speeds, accidents will happen, which sometimes results in death. A completely computer controlled vehicle operating on an elevated roadway will not cause these deaths. It will eliminate virtually all of the Interstate deaths caused by accidents.[22]

---

22   Sholnn Freeman, "Title??," Washington Post, August 23, 2006, http://www.washingtonpost.com/wp-dyn/content/article/2006/08/22/AR2006082201152.html

*I like to drive my Ferrari 200 mph through the desert.
I want to go as fast through cities, that shouldn't hurt.*

Chapter 12:
Human Actions While Driving

Today good driving is heavily dependent on courtesy, which all too often seems lacking. Too many people demand getting their way on the roadways because of the high value placed on time. Conserving time is important, people shouldn't have to spend as much time traveling as we do today. Unfortunately very many people seem to think they own the road on which they travel. Some of this apparent lack of courtesy may come from an unconscious thinking that since they pay taxes, which help to pay for road maintenance and upgrades, they sort of think of themselves as part owners of "their" road. Need for road courtesy just won't be a problem on future computer controlled roadways, when the ultimately courteous HEVER computer systems will be designed, built, and implemented.

Currently persons speed on all kinds of roadways. Sometimes they slip by without getting stopped, sometimes they get tickets, which can get costly and even cause loss of their driving privilege. Note that driving is a privilege, not a right; in the future, more control by computers will eventually make this a reality. Travel in a totally automated world will not need tickets, now caused by such things as speeding or freight overloads. Future computer controlled systems, by their nature, will obviate any such need for tickets. This will improve the transport experience and should free up some police time for more serious law enforcement.

On today's roadways, drivers often drive for miles behind another slower vehicle. Due to wanting to avoid remaining stuck behind such a vehicle, another vehicle, one not immediately behind the slow vehicle sometimes decides to pass. The driver originally following immediately behind the leading slow driver often will pull out, cutting off the vehicle behind him that had begun to pass. This prevents that vehicle from passing and creates a dangerous situation. If the driver of the lead passing vehicle goes barely faster or about the same speed as the original slow vehicle, this causes the cut off vehicle to slow down abruptly. On future computer control roadways, drivers of vehicles will not experience this frustrating and irritating situation.

Many people basically fear high speeds when they perceive a significant potential for accidents. If an accident occurs, the probability of death increases at higher speeds. Human beings marvel at the ability of today's automobiles to go 40, 50, or even 100 miles per hour. When one considers that today's vehicle, controlled by imperfect humans who use vehicles virtually unconstrained in any horizontal directions in which they can traverse, one can see that accidents can easily happen. Sometimes vehicles even go radically airborne, totally out of control. During heavy rains, automobiles can hydroplane at about 40 mph. This means that the tires are on water and are not actually touching the roadway surface. This same unconstrained hydroplaning phenomenon makes vehicles more dangerous as speed increases.

Ability to avoid causing accidents involves alertness and increased driving ability when attaining higher speeds. Racing vehicles can normally go 200 mph or more. Commercial jet airplanes can fly as much as 500 or 600 miles per hour. Most of our current automobile operation occurs in a relatively unconstrained manner. The high speed phenomenon takes away some of the ability of completely controlling the vehicle for accident avoidance, essentially taking control out of the hands of the good or bad driver. One often cannot avoid accidents where someone else makes a mistake, does not drive properly, or a vehicle fails mechanically, and in some manner, an accident happens. All too often, the person doing good driving becomes an unwary victim in accidents. And sometimes drivers don't cause an accident, it just happens.

Who cares if roads are built to wear out, why do we have fears?
The Roman Apian Way is still around, what a waste for 2,000 years.

Chapter 13:
The Future

The Hi-speed Electric Vehicle Elevated Roadway (HEVER) system will have a completely computer controlled elevated roadway system. Automobiles, trucks, and buses will operate at any time of night or day.

In a joint effort, computers of the HEVER system and vehicles will monitor and control vehicles while they are using the system's roadway. The HEVER system, using electromagnetic propulsion and highly computer controlled vehicles will transport material, products, and people at relatively high speed while on the system and the vehicles will operate non-stop. The vehicles will move at relatively low speed while traveling off of the HEVER system.

The HEVER system will complement the existing Interstate system and will have a design goal in which the system will last at least one hundred years. As a safer, more efficient system, it will complement the current Interstate road system. With some exceptions, due to practical limitations, this future roadway system, will handle most of the current types of Interstate transport of people, products, and materials. As examples, obviously open truck beds and soft-top automobile convertibles won't be practical on a high-speed roadway; dual trailers like those on current eighteen wheeler trucks won't provide adequate safety on a high speed roadway. Even with limitations placed by setting safety and other standards, the development of a high speed, highly computer controlled roadway, will not just provide nice benefits to our society. While we can walk away from this opportunity, we must create the HEVER system roadway if we wish to save lives. Its development will also reduce pollution and save valuable oil resources for other uses.

By considering the use of pillars with relatively small footprints, development will require a far smaller amount of land than we currently use for creating today's Interstate roadways. Constructed pillars using concrete with metal reinforcement will hold the roadway in place, relatively high above the ground. Adequate peering, as deep as necessary, into the ground will add support to the roadway system. Development will require significant design. Newer, higher quality materials now exist that will resist wear and tear. Every day usage of today's existing roadway materials does not adequately provide resistance to wear and tear.

One hundred bunkers, built for an ordinance company in the 1940s at a facility near Weldon Springs, MO, originally held explosives—TNT and DNT. The explosives were manufactured for the military to use in World War II. Of the original one hundred bunkers seventy-six, presumably too contaminated to leave open, have been sealed while fourteen still remain open. Prior to and during World War II, engineers, working for the military, had these bunkers built using concrete that made the bunkers almost indestructible. Use of this type of concrete on an elevated roadway would provide dramatically improved longevity and lower maintenance cost to a future elevated roadway. The concrete, too difficult and expensive to destroy, resulted as the reason for sealing the bunkers. Built for wartime, these bunkers were obviously constructed to last, and even today would have the capability of resisting most bomb attacks. This concrete is harder and doesn't wear anywhere near as much as most of our commercial concrete in use today.

Our roads today deteriorate quite easily, requiring significant repairs of potholes and sometimes resurfacing or complete replacement. They are built considering the cost to build versus amount of time that they will exist without needing repair or replacement.[23]

An article stated that the Weldon springs TNT plant covered 18,000 acres. After World War II, much of the property was

---

23   "Environmental Pathways, Environmental Contamination and Other Hazards," http://www.atsdr.cdc.gov/HAC/pha/wel_sp/wsta_p2.html.

released to the state of Missouri. Developers used part of the land to create the Busch Wildlife Area. The remaining TNT plant became a uranium processing plant for a while. There still is a large containment bunker that remains for the purpose of storage of hazardous waste.[24]

An Internet article dated October 22, 2009 discussed geopolymer material, which it described as synthetic aluminosilicate material or super cement, like a ceramic that doesn't require firing. Scientists have now proven that in ancient Egypt, builders basically manufactured the huge building blocks by pouring this type of cement in place; they now believe the pyramid blocks not to be stones.

At the times of the Romans, builders used crushed rock mixed with burnt limestone as a building material. This type of material, unreinforced, was used to build the dome on the Pantheon in Rome, which remains just as strong as when the Romans built it about two thousand years ago.

A Ukrainian scientist, Victor Glukovsky, discovered that by adding alkaline activators to the ancient cement of the time, an improved more durable material resulted. The new kind of cement yielded significantly higher quality and longer lasting material. Glukovsky's work lead Joseph Davidovits, a French chemical engineer to discover the chemistry behind geopolymers and their use. Professor Davidovits determined that Egyptians used poured cement rather than stone for the material of the Egyptian pyramid building blocks. Workers basically manufactured the blocks by using cement that ended up looking like stone. While in a liquid form, they apparently poured the blocks in place rather than lug large stones up the pyramid steps. Scientific X-ray techniques have supported the contention that the Egyptian pyramid builders did not use stones but actually used manufactured large blocks. Apparently, the construction of nuclear plants uses extra hard concrete. Any future elevated roadway should consider use of extra hard concrete.[25]

---

24   "Weldon Springs TNT Plant," 1/27/2008, http://www.waymarking.com/waymarks/WM3204_Weldon_Springs_TNT_Plant.
25   David Hambling, "Super Concrete in the U.S. Military, Iran...and the Pyramids?" October 22, 2009, http://www.wired.com/dangerroom/2009/10/super-concrete-in-the-us-military-iran-and-the-pyramids/.

HEVER

No one should consider this discussion of hardened concrete as anything but an example for consideration among some other construction alternatives. Nor should anyone consider it as a final design, it is merely intended as an example, an alternative to consider.

Design of these new roadways will resist earthquakes. The HEVER system will have pillars that will hold the roadway relatively high above ground. These pillars might have a base that will allow the pillars to use sliding mechanisms to move small distances along the ground. The roadways on top of the pillars might also have the ability to use sliding mechanisms to move across the top of the pillars. These roadways themselves could allow limited telescoping of the roadway in such a way that vehicle drivers could continue safely on their way even should some minimal shifting or minor earthquakes occur. This shifting movement will allow the HEVER system to remain intact as long as the pillars and roadways maintain integrity.

Virtually no structures can avoid damage from all earthquakes. The HEVER system's roadways will resist negative effects of some minor earthquakes and its design will allow for fast and relatively easy repair and/or maintenance. Since a goal of its design will dictate the use of standard interchangeable parts, the HEVER system will thus have the ability to minimize time for making repairs, even if the earthquake occurs in a highly disruptive manner. As such, it, therefore, should provide relatively lower costs to its travelers. Always, the HEVER system will use high-quality, standardized, interchangeable parts as much as possible.

All HEVER system roadways will be elevated above the ground. In some cases, this will even yield a less expensive cost than that for moving vast amounts of dirt to create a roadbed for the eventual covering with concrete or blacktop. Often, construction personnel can do much of the work while using prefabricated components. They can use finished portions of roadways to deliver materials to extend roadways further. The structure holding the roadway in place will have a relatively small footprint. The existing Interstate system will require less adding of lanes. Due to the development of the HEVER system, this new roadway will also cause

THE FUTURE

a reduction in the need to acquire and use additional amounts of land for expansion of Interstate highways in the future. Less additional land will be covered with concrete.

Currently, in some areas, especially in high population areas, too much green space currently gets covered with concrete, sometimes even changing the water flow and amount of water runoff. In some areas of our country, this now sometimes causes flash flooding that never happened years ago. The construction of the HEVER system will require use of less green space than that now required for constructing concrete highways. In the past, developers have tended to cover Interstate road beds with concrete. Since the HEVER system will use less green space than that used for Interstate road beds, problem areas where flash flooding caused by water flow changes will not increase as rapidly.

Compared to designing for non-elevated road beds, it easily comes to one's attention that elevated roadways provide considerably more practical design alternatives. The designs for roads crossing other roads must result in easier and more straightforward designs to and from the HEVER system roadways. Having elevated roadways makes certain that the system will provide more safety for pedestrian traffic than in the past. The elevated roadways create an environment conducive to having an integrated computer control system and a closed computer communication network for performing the control. All of this makes better and safer traffic control more possible.

In the past, railroad companies, due to high cost considerations and political obstacles, avoided elevated rails, even through congested areas. If railways had originally been elevated, trains could have gone much faster. Today they would most likely compete more successfully for the business they lost to the trucking industry. In the future, it will prove more effective to produce a system of elevated roadways that better serves individuals by allowing automobiles, trucks, and buses to operate on or off of such a system and while on the system, operate without multiple loading and unloading of freight or luggage from start to finish of a journey.

The HEVER system of roadways will exist as a reasonably high, elevated structure. Developers will build the roadway high enough to give adequate safety; they will create as straight and level of a roadway surface as possible. We can never accomplish a proper amount of safety for travelers and pedestrians by having roadways built with ground-based road beds like those of today's Interstates. Too many people and vehicles need to use the ground level for transportation. The elevated HEVER system of roadways will not add to occurrence of non-roadway accidents. Crossing the roadway will only allow movement above or below by people, vehicles, or other roadways. Because of computer control, merging vehicle traffic to and from separate roadways will safely cause automated yielding as necessary. Computer system control will virtually give an assurance of safe collision avoidance. Pedestrians will enjoy a safer situation because of elevated traffic existing above them. The HEVER system lanes will provide a better experience since its higher speed will not endanger pedestrians and its elevated roadways will provide safer vehicular accesses to and from other merging HEVER system roadways.

Variable speeds can cause drivers of vehicles to create irritating, inefficient, and sometimes dangerous driving situations for other drivers. For example, some drivers increase the speed of their vehicle in order to get even with cars and then slow down as they pass and continue to go slower after passing. Sometimes, especially when traffic is heavy, this can cause traffic blockages to occur. Even today, drivers cannot easily use the convenience and safety of cruise control in these situations.

The future will bring significantly better design than in the past. We will make this type of roadway completely automated; no other known roadway design can come close to matching the benefits of its design. The HEVER system's computer control will always control speed making driver assist features impractical. Even today at high volume situations, "assist," such as automobile cruise control cannot operate without great difficulty.

To be safer and have more potential for efficiency, developers will build future elevated roadways that will have safe access and complete computer control.

*Sensors, sensors, sensors—there's a lot of duplication.
Is this what we want, it seems like lots of repetition.*

Chapter 14:
Switches, Brakes, and Sensors

All entries to and exits from the HEVER system must have a "Y" type branch with no complicated entries and exits (such as cloverleaf arrangements). Also, "Y" intersections will exist as the only choice for directional change that the HEVER automated transport system will require and allow. The "Y" intersection would start out by creating a very narrow separation. The choice for entry or exit from the HEVER system will always be on the far right side of the HEVER system roadway used by drivers while their vehicles travel on the HEVER system. Inadequate for future vehicle travel on mass transport systems, designers must avoid mechanical switching systems. If any switch failure situation can occur, it should fail in the safest direction.

The HEVER central computer system will work together with the vehicle computers in order to determine the lane that the vehicle must take. Electromagnetic devices will cause the lane switching to happen when the need to change lanes arises. It will happen according to the predetermined routing in the vehicle's computer system. Vehicles do not actually have a straight through and a right or left direction choice. When a vehicle comes to a place where the roadway comes to a "Y" intersection (switch), the vehicle must take one path or the other. The central and vehicle computers control electromagnetic propulsion devices that pull the vehicle either toward the right or left lane. Because of the predetermined routing information, "Y" switching occurs and the vehicles automatically take either the right or left path.

Mechanical switching does not actually takes place. The electromagnetic propulsion system will essentially use the central computer and the vehicle's computer to control the vehicle by pulling it right or left. Situations like this happen when vehicles need to change lanes or use an exit when traveling on the HEVER system. This provides an additional benefit inherent in this type of switching system; it provides two computer systems that while they usually act by making joint decisions, if one fails, the other can act independently. When necessary, one computer functionally backs up the other in a case where one of them fails. Assuming one of the computer systems functions properly, the failure situation is referred to as "fail soft."

Many advantages of electromagnetic-type switching and braking systems make them a better choice over mechanical-type switching or braking. Mechanical switching will eventually wear out because through its normal usage, friction causes wear. There is virtually no wear and, thus, greater reliability when using electromagnetic devices. Electromagnetic devices will have the ability to function better under some failure situations (graceful degradation). Computer systems will provide diagnostics indicating vehicle status. They will make use of electromagnetic devices to interpret safety conditions indicated by devices use of the electronic sensors. The computer systems will determine the status of vehicle conditions much faster and give more accurate results for electromagnetic devices than for mechanical, friction-based devices.

Manufacture, installation, and maintenance of electromagnetic brake and sensor devices will generally provide a lower cost than that for friction based brake systems. Bad weather and other wear and tear phenomena will negatively impact operation of mechanical and friction based devices. Magnetic devices will provide easier, more accurate, and better monitoring and controlling by computer.

Future electrically powered vehicles will use a great number of very low cost, very powerful sensors and computers. Vehicles will employ multiple sensor readings that essentially will vote. This will allow vehicles to make the determination of which action

SWITCHES, BRAKES, AND SENSORS

to take. The vehicle and the central computers will receive sensor information and will use it to monitor and control the vehicle and the entire HEVER system.

The central computer communication system will keep track of the vehicle's location at all times. Some information, originally gathered by vehicles using computers and sensors on board the vehicle itself will mostly be transmitted to and, as necessary, will reside in the HEVER central computer system's database. Many sensors will exist on the roadway. These sensors will alert the vehicle's computer system that a switch situation ahead will require the vehicle's computer to make a directional decision. The vehicle's computer system will execute instructions as dictated by routing information held in the central and vehicle's computer systems. The vehicle will execute and make the necessary directional change in plenty of time.

The vehicle's computer will create or gather information, some of which will exist on the vehicle's computer system. All necessary information will also exist on the database of the HEVER central computer system. Most of the data from all of the vehicles operating on the system roadway (some of it used to produce statistical reports) will reside on the central computer system's large database. Some information asked for by travelers will come from the central computer system. The information that the vehicle's computer system will send to the vehicle's computer input/output monitor will, for example, consist of descriptive information about the vehicle's status, vehicle components, or data entered or changed by the traveler.

The central computer system will, at all times, know the exact status and location of all vehicles so that the overall HEVER system's computer can ensure collision avoidance. Both the vehicles' computers and the central computer system will act to make sure that vehicles safely merge into lanes and, as such, have enough space.

*Wheels, motors, and parking, this I understand, but spurs.
What is this, some kind of cowboy curse?*

Chapter 15:
Wheels, Motors, Spurs, and Parking

Vehicles will not require the use of wheels while vehicles use the HEVER system. All vehicles will need sets of wheels and friction brakes prior to entering or when leaving the HEVER system. While on the HEVER system, vehicles will not use their motor and drive systems, which they use while the vehicles operate off of the HEVER system. No vehicles' batteries will provide power to vehicles while the vehicles operate on the system's roadway. Rather the system will transport temporarily non-powered, vehicles in the same manner that vehicles' contents will travel the roadway. The HEVER system will, so to speak, transport vehicles as freight.

Vehicle wheels will need to operate as soon as the vehicle leaves the HEVER system and all the while vehicles travel off of the system roadway. Vehicle wheels will also need to operate until the vehicle gets up enough speed for the electromagnetic propulsion to take over. After entering the roadway, the wheels' motor will no longer power the vehicle; the electromagnetic propulsion will have taken over the transporting of the vehicle.

While not likely as a design, if necessary, the wheels will rise to an appropriate height and will eventually stop operating while the vehicle travels on the system roadway. Needing wheels to rise and lower will depend on the type of electromagnetic levitation and propulsion chosen. If the design requires it, the vehicles will raise the wheels somewhat like landing gears on airplanes. Therefore, vehicles will not operate their wheels while on the HEVER system. When driving off of the HEVER system, the vehicle will use wheels that should have relatively high performance standard tires.

HEVER

Other than the possible raising and lowering of vehicles' wheel systems, vehicles will not use the vehicles' power while they continue to travel on the HEVER system. The roadway design will most likely not require a lifting of the wheels. A roadway elevated high enough and made up of flat metal surfaces used to levitate vehicles, will most likely allow the wheels of vehicles to, so to speak, hang down naturally, or basically dangle, while vehicles travel on the system roadway.

Leaving the HEVER system, vehicles will operate similar to their entry process but in a somewhat reverse manner. Vehicles will require usage of a friction brake system for driving while not on the HEVER system.

Many manufacturing, distribution facilities, and other facilities (such as large housing developments) will have the option of using spurs as part of the HEVER system. Spurs will allow a more automated and efficient delivery of material and products to destinations and from points of origin. Since vehicles will have the ability to be transported automatically, spurs will have the ability to also allow a more automated and efficient delivery of new trucks, buses, and automobiles to facilities at or close to dealers. Spurs will allow a more automated and efficient return of empty trucks to sources. The HEVER system will require truck drivers (at least for loading and unloading or for security) for driving after leaving the HEVER system. Where spurs have appropriate computerized controls and adequate automated parking, the system will allow unmanned vehicles such as a return of empty trucks.

Using spurs will require an increase in speed when entering and a decrease in speed when leaving the HEVER system. Spurs will require enough power to quickly enough reach an adequate speed and headway while entering onto the system roadway. Headway consists of the time interval between two vehicles traveling in the same direction in the same lane on the same roadway.[26] Enough speed must allow the vehicle to no longer require use of wheels and to provide the ability to control the inertial forces so that vehicles can safely enter onto the HEVER system roadway.

---

26   Definition using Merriam-Webster's Online Dictionary, Retrieved from www.Google.com/Dictionary

WHEELS, MOTORS, SPURS, AND PARKING

The HEVER system will need adequate distance on the HEVER system roadway for slowing down and entering an exit pathway.

There must be short-term parking space available, if needed, for temporarily parking of the vehicle. Should there be an individual who can drive the vehicle off of the parking area, the vehicle driver will not require use of the parking space, but it must be available until the vehicle leaves the (potentially needed) parking area. On and off spurs can locate virtually anywhere on the HEVER system as long as they can accomplish standard safety checks, using standard, approved vehicle checking equipment, which must exist at all spur locations. Available personnel must handle the spur facility twenty-four hours a day, seven days a week. There must be enough space to park vehicles; vehicles left on the parking lot too long will be removed to a more remote parking facility. There must be enough space (land area) for a long exit lane in order to guarantee the ability to reduce speed safely at off ramps and for accommodating adequate temporary parking.

*We should build a roadway system solely using the lowest cost activity.
Unless first and foremost safety needs to have the highest priority.*

Chapter 16:
Some HEVER System Benefits

The HEVER system will make it practical for people to live anywhere desirable and yet have fairly quick access to facilities or locations considerably distant from their homes. This type of travel, in terms of time, will provide closer destinations than ever before for places where people work, live, shop, or recreate. The HEVER system will cause an increase in the potential for the development of new concentrations of locations for living, playing, shopping, and working. This will, in turn, make for a more efficient and enjoyable life. After development, these centers certainly will continue to grow much larger, contain more products and jobs, while providing far more convenience than we now experience. This will often cause apartment living, shopping centers, entertainment establishments and business offices, commercial, industrial, and educational facilities to exist increasingly in the same proximity of each other. Educational administrators might find it practical to locate one or more universities at these highly convenient locations in cities. In general, this would create an environment very convenient for people.

The HEVER system will offer one of the best opportunities for cities to attract population growth due to the concentration and development (growth) of business, shopping, and entertainment. Of course, managers of cities should continue to clean up the current bad areas of cities. While not a panacea for city rejuvenation, the development of the HEVER system will help cities to experience some renewal due in part to this ability for people to rapidly access facilities in or near their cities as well as other cities.

HEVER

Also people choosing to live in cities would enjoy computer controlled, high speed, and personalized access to other cities and to rural areas. They could relax, sleep, or do work during trips on the HEVER system and they could take some belongings with them and use their own vehicle after reaching their destination.

Vehicle accidents will virtually not happen while vehicles operate on the HEVER system. Design of vehicles and components will include first and foremost safety as the highest priority. Design will include good aerodynamics and to some extent, weight and size of vehicles.

Electric automobiles will have sizes like today's automobiles, at least as long as or longer than current automobiles. If long enough they will likely be called limos (buses?). Trucks will come designed in maximum sizes comparable to the current eighteen wheeler truck lengths and widths. The major design need of how to handle inertial qualities will cause the imposing of vehicle size and weight limits, which may or may not have similar limits like those currently in use. Automobile and truck sizes will have standards mostly designed considering today's typical usages for transporting freight and travelers. The vehicle length, width, and height will vary but vehicles will have limits. The number of seats and optional internal makeup will look similar to that now in usage, hopefully with considerable improvements. Whatever limits developers decide to place on vehicles, the HEVER system will enforce them when vehicles attempt to enter the system.

All vehicles using the HEVER system will have computer systems capable of checking sensors of said vehicles for potential problems. Kiosks' computer systems and the vehicles' computer systems will safety check the vehicles before entering and after leaving the HEVER system. All computer systems will perform some vehicle related functions while vehicles travel the system. While vehicles operate on the HEVER system, their computers will do constant monitoring and safety checking. The vehicles will use their on board computers to act in cooperation with the HEVER central computer system, both constantly monitoring and checking in order to look for potential problems. Vehicle computer systems and/or the central HEVER system's computer can determine

SOME HEVER SYSTEM BENEFITS

and cause vehicles to leave the system roadway and use alternate routings.

The HEVER system will provide electric power to vehicles and will power the electromagnetic propulsion devices needed to transport them. The HEVER system will also provide electric power to all of the vehicles' equipment, including the vehicles' computers. If the vehicle's batteries have a somewhat low charge when the vehicle enters the HEVER system and assuming that time allows, the HEVER system will bring the vehicle's batteries up, as close as possible, to a full charge.

Currently, virtually endless potential ways to generate electricity exist. The electricity production industry uses processes and systems now in place to provide vast amounts of electricity. New methods will allow the ability to provide enough electricity production to meet any increase in demand and at an ever decreasing cost. Electricity is a virtually unlimited resource; we just have to produce it without causing waste and pollution. Where possible, the HEVER system will use solar cells for electricity generation.

A battery system will reside on all vehicles. While vehicles do not need to use their batteries when traveling on the HEVER system, they will use their battery systems to power their vehicles before and after system usage.

Many individuals and industries will benefit from job creation. The system will cause creation of jobs, bringing economic benefits. Costs will reduce for manufacturers and for the some other users. Perishable products will move faster and at a lower cost. Consumption of fossil fuels will significantly decrease. The military will enjoy lower cost and faster movement of men and materials.

Vehicles will use the HEVER system at considerably higher speeds than those experienced on today's Interstate. While the HEVER system transports people in a (temporarily) non-powered electric vehicle (while on the system), travelers will experience going at a speed somewhat below the normal maximum speed for the fastest commercial passenger airplanes. Even with the attaining of high speed, accidents caused by human drivers just won't happen due to advanced central computer control, adequate

HEVER

vehicle computer control, sensor redundancy overkill, cross sensor monitoring to predict potential problems, and checking vehicles prior to their entry onto the HEVER system.

As a norm, vehicles will operate at higher speeds, due to taking control of vehicles out of the hands of humans and by constraining the vehicle on a roadway while on the HEVER system. Considering vehicles that will use such a high-speed system as the HEVER system, the vehicles will require a most careful and thorough design, especially for safety purposes. The vehicles must carry a practical load and the roadway must handle the load and travel at relatively high speeds. Designers must recognize the importance of establishing appropriate limits on weight, height, width and length, especially due to need for design that will provide safety and more than adequate control of inertial issues. The determination and setting of appropriate inertial limits for trucks and buses will cause a need for more significant design than that for similar associated automobile inertial limits. The design of trucks and buses will require extremely high standards and will not allow transport of some products due to size, weight, or the dangerous nature of some products.

Raised significantly above the ground and allowing only completely electric powered vehicles to use it, the HEVER system will completely control all transportation on the system by computer. Drivers of vehicles will still use existing Interstate roadways when necessary or desirable.

The HEVER system will not allow use of vehicles with freight containers not integrated with the truck cabin. Rather all vehicles on the HEVER system will consist of a self contained, somewhat self controllable, singular unit, mostly controlled by its own computer system acting on a cooperative basis with a large central computer control system.

Testing, one, two, three, four... testing, one, two, three, four. Why test so much? Fix it at the dealer's store.

Chapter 17:
Testing

The building of a complete testing facility will happen after determining short- and long-range testing requirements.

For the first test facility, construction personnel will build a relatively short section of the actual HEVER system's elevated roadway. Developers will build this initial section with as low an elevation as possible yet high enough to meet all future minimum height requirements. Appropriate portable raised platforms will provide working space for temporary testing facilities. Additional production roadway construction will continue during the first phase of testing. After adding future constructed sections of roadway, the testing capabilities will expand. Test vehicles will use this first section of the future HEVER system roadway for some initial testing to get the electromagnetic levitation devices working properly. Following that, testing will occur for the electromagnetic propulsion devices. In the beginning of the implementation of the temporary test environment, personnel will implement the roadway system and supporting central computer systems. As the roadway is lengthened, workers will build a parallel, portable roadway with at least two electronic "Y" switches.

When starting the roadway and vehicle testing, the system will use prototype truck and automobiles. Humans will not occupy them in any of this initial testing.

Production of a relatively finished prototype truck(s) will contain as much full function as possible and it will act as a test vehicle(s). Testing personnel will use prototype vehicles on the finished stretch of the HEVER system roadway to work on as many safety issues as possible. The test system will especially evaluate speed and load limits needed to satisfy inertial safety requirements. The test vehicles will operate empty and, when practical, will alternately carry a full load. The vehicle(s) will operate at varying speeds. Electromagnetic propulsion can ramp vehicle speed up very quickly to almost unbelievably high speeds. When rigorous testing indicates that the computers, roadway, prototype vehicles, sensors, and devices have all functioned properly at lower speeds, the test system can allow testing of vehicles at increasingly higher speeds. When doing the above testing, the vehicle's computer system (and possibly the central computer system) will need to have availability to the test environment in order to accomplish increasingly advanced capability. At this initial testing, during execution of any test, prototype test trucks will not have any humans on board.

After the successful testing of trucks, test automobiles will use the HEVER system roadway and testing will happen in a similar manner to that for trucks. The duration of this test period will depend on the ability of the test automobiles to pass all testing requirements. During this phase of the testing of automobiles, no trucks will operate on the test system and no test automobiles will have humans on board.

Eventually to prove success, both trucks and automobiles will operate using regression testing procedures in a manner similar to previous tests. Regression testing implies that as future tests occur, testers must also include employment of past test criteria. After initial regression testing successfully completes, testing will then allow addition of some new, additional test procedures. Around this time, construction personnel will have built additional temporary sections of test roadway for the purpose of testing some of the initial collision avoidance software. These testing sections, built for testing collision avoidance, and temporary in nature, could eventually move to a more permanent test facility. More likely, its reuse will cause a need to move it and use it to further extend the construction of the actual HEVER system roadway.

The long-term testing facility most likely will not require an elevated roadway; logistics of having the long term test facility roadway on the ground will likely prove more beneficial to testing equipment and personnel. Determination of whether to build an elevated or ground-based test roadway (or both) for the long-range testing facility can happen while initial testing occurs on the temporary system. The subsequent construction of the long range can also occur in parallel to building and using the temporary test facility that will eventually find use as part of the production HEVER system roadway.

Testing, gathering of statistics and simulation studies will continue while building a full roadway between at least two major cities with intermediate entries and exits at smaller cities. By this time, completion of any entries and exits planned for this production roadway must have occurred so that testers can begin pre-production testing. These entry and exit locations must have kiosks with the ability to check for vehicle problems. Personnel must be in place to handle the kiosks and vehicles making entries to the roadway or exits from the roadway to a parking area.

After doing and redoing testing of a completed roadway between two cities, full implementation will happen. Trucks will implement before automobiles. Only fully automated vehicles with no humans on board will initially operate on the roadway for a limited time. Manufacturing facilities can employ spurs for entry to and exit from HEVER system locations. New trucks and automobiles, without drivers and fresh off production lines, could be sent from the manufacturer to distribution points.

Eventually the HEVER system will employ trucks with humans on board. Testing will also continue through use of the long-term testing facility.

The HEVER system will allow the addition of vehicles that will eventually transport humans in a similar fashion to that of traveling in a limo or a bus. For sizable buses to travel on the HEVER system, they will have a design in length, width, and height limits similar probably to that for automobiles. In actuality, buses will likely be very long limos.

*Who needs backup equipment for vehicle electric batteries?
Backup systems cost money; spend the cash; don't be a sleaze.*

Chapter 18:
The HEVER system

The HEVER system will eventually implement usage of the roadway by automobiles containing humans. At this point, the system will control these vehicles by computer and the vehicles will use only electrical power. In some cases, vehicles will have a back-up gas generator to produce electricity for emergency use when operating off of the HEVER system roadway.

It is anticipated that one high-speed lane could handle at least seven and perhaps as much as eight times the volume of one of today's lanes. This assumes an ability to travel 400 mph where vehicles operate in a non-stop manner from the time they enter the HEVER system and until reaching their final destination. Assuming vehicles travel at 400 mph for the HEVER system and an average speed of 60 mph on the Interstate, a new single lane would handle 6.67 times that for the typical Interstate lane. Obtaining 400 mph on the new HEVER system is more likely than averaging 60 mph on our Interstate lanes.

More people will likely bring food for a meal during a HEVER trip. More slowdowns and stops will likely happen on our Interstate lanes. Accidents, gaper blocks and other slowdowns or stops just won't likely happen on the new roadway. Capacity should not present itself as a problem since a minimum of two lanes in each direction is anticipated.

At least two lanes will exist in each of both directions on the roadway system, and the roadway system design will allow the addition of more roadway lanes to the left of the aforementioned lanes. Only the leftmost lane(s) will allow the highest maximum speed and will normally handle all vehicles, trucks, buses, and automobiles. Through cities, developers will build one more extra high-speed lane (a total of three or more lanes) to accommodate rush hour or other high volume traffic or for use as a lane for preparing to exit. The rightmost lane will have the slowest maximum speed. The HEVER system will accommodate lane changes from higher speed (toward the left) lane(s) causing reduction in speed if moving toward the rightmost lane. This will allow controlled, safe entry to or exit from the HEVER system roadway. Entry and exit is a significant purpose for the rightmost lane. Vehicles using lane(s) to the left of the farthest right hand lane can move to the right toward and eventually into the rightmost lane in order to slow down and exit. All vehicles will also use the somewhat slow rightmost lane for entry. Allowing enough slow down roadway and temporary parking will present a significant issue for designers.

Requirements when transporting trucks and automobiles on the HEVER system will require quite significant design efforts in order to build a highly computer controlled roadway emphasizing ease of getting into, out of, and through a city. Due to the high concentration of business and commerce in cities and large populations in the suburbs, higher traffic volume flows into cities in the morning and away from cities in the evening. In general, the Interstate system uses land as efficiently as possible. Just as the Interstate has handled traffic in its past, it should continue to operate as such in the future. They will use it in the foreseeable future, driving during times of lower volume, and they will operate at lower speeds. With the development of the HEVER system, driving the Interstate in the future will give a better driving experience than it does today. In the future, development of the HEVER system will obviate the need for much of the otherwise construction and growth of our current Interstate system and will even result in lower Interstate system day to day maintenance costs.

One problem faces society, that of an almost infinite number of opinions (especially political and special interest groups) as to how the cities' road systems should grow.

In the future, construction personnel will increasingly do much of the work using standard and mostly prefabricated com-

THE HEVER SYSTEM

ponents. They will use finished portions of roadways to deliver materials, thereby, further extending roadways. In the future, the development of the HEVER system will also cause a decrease in requests asking for additional land for Interstate highway right of ways. There should be almost a zero need for land requests for the Interstate system.

Currently, in some areas, especially in high population areas, construction covers too much green space with concrete. Sometimes it even changes the water flow and amount of water runoff and causes some flash flooding. The HEVER system will use less green space than now required for concrete and blacktop highways. By using less green space and not covering ground with concrete or blacktop, problems of flash flooding caused by water flow changes will not increase as rapidly. The HEVER system elevated roadways will not have a problem clearing snow and ice from roadways. More design alternatives are available with raised roadways, especially in handling roads crossing other roads. Pedestrian traffic will be safer (virtually not an issue). All of this makes better traffic control more possible.

In the past, railroad companies, due to high cost considerations and political obstacles, avoided elevated rails even through congested areas. If railroads had been elevated, train speeds would not have been as artificially limited and trains could have gone much faster. If train speeds had not been so restrained in the past, today's trains would likely have a much greater ability to operate competitively. Today it will be more effective to produce a system of elevated roadways that better serves individuals and their vehicles, vehicles that can operate on or off of such a system.

The HEVER system will have reasonably high elevated roadways, high enough to give adequate safety. Builders will need to create as straight and as level of a roadway surface as possible. We would never accomplish a proper amount of safety by having roadways built with ground-based road beds like those of today's Interstates. With the elevated HEVER system of roadways, accidents virtually will not exist since crossing HEVER system tracks would exist above or below each other. Merging traffic to and from roadways would yield to decisions made by computer control

HEVER

more easily. Pedestrians will enjoy a safer situation because of elevated traffic above them. The HEVER system lanes will provide a better travel experience since the higher speed will not endanger pedestrians and will provide better vehicular accesses to and from crossing HEVER system roadways.

The future will bring significantly better design than in the past. We can make this type of roadway completely automated; no other roadway design can come close. The HEVER system's computer control will always control speed, making driver assist features impractical while using the system. Even today in high volume traffic situations, assist functions such as automobile cruise control, cannot operate without great difficulty. To be safer and have more potential for efficiency, future highway roadways must be elevated, have safe access, and be highly computer controlled.

*I cannot see the road ahead of this eighteen wheeler truck.
I know I can pass; I know I can pass; all I need is a little luck.*

Chapter 19:
Trucks

The first implementation of the HEVER system will only include trucks. However, future design must allow use by automobile and bus-type vehicles for human transport.

The HEVER system truck will operate somewhat like present trucks in that it will carry relatively large freight loads and will operate on virtually any of the nation's current road systems. Trucks will employ designs that will optimize aerodynamics. They must adequately handle highly varying inertial issues. Unloaded trucks must have a weight that as light weight as possible. The materials used in the construction of trucks should include use of large amounts of plastic and aluminum.

Trucks will operate using power from electric batteries when off of the HEVER system. When the truck travels on the HEVER system, it uses the system's electromagnetic propulsion to transport the truck as if it were freight. Appropriately, the truck's motor and drive system do not need to operate. While trucks are on the HEVER system, no power comes from the truck's batteries, the truck does not actually operate its electric motor. While operating off of the system, the truck will use that battery powered electric motor to power the truck. In some cases, practicality may require having an on board gas generator for backup purposes.

Charging of the trucks' batteries will occur to the extent possible while on HEVER system. As much as possible, and if practical, trucks will use solar cells which will help to save battery power for use while they operate off of the HEVER system.

HEVER

Solar collectors located along the roadway could offer an alternative for providing electrical energy.

If electricity generation from hydrogen fuel plants proves viable and non-polluting, the HEVER system might use them located strategically along its system roadway. The system will have the ability to use furnaces to generate electricity, burning replaceable non-polluting resources such as garbage, weeds, and corn. Even coal could be used if scrubbing can eliminate pollution. The HEVER system will make use of current electric utilities. The HEVER system's energy producing facilities might create excess electrical energy. It will then make available surplus energy produced by the system available for use by electric utilities at a little or no cost to the utilities.

Interior and exterior, especially for the cab area, will emphasize protection for drivers and passengers. The main truck cab interior will have seats like the trucks of today, hopefully with many improvements. The interior will emphasize comfort for and efficient use by drivers and passengers and will have space for clothes and other necessities. To the extent that weight and other issues allow, trucks will have telephone, fax, small office facility, television, radio, computer entertainment, and lavatory, as options. They will have these and many other new options due to highly available low cost electricity.

Interior and exterior of the portion of the truck used for carrying freight will require some standardization and will, to the extent possible, accommodate current normal trucking freight package sizes.

Truck wheels used when drivers operate off of HEVER system roadways will likely have solid tires (probably rubber) rather than inflated tires. Trucks will have a standard friction based braking system for use when not operating on the HEVER system.

While on the HEVER system, trucks will operate at high speed. Truck drivers will operate their trucks at a slower speed when not driving on the HEVER system.

Truck computer systems will automatically monitor the truck and gather data for statistical purposes. The computer system on each truck will look for problems. The truck's computer system will also communicate with the HEVER central computer system to see if its computer system, using roadway-installed sensors, detects any problems related to the truck or the roadway. The HEVER system will not allow trucks onto its roadway if its computer system or the central computer systems find any potential truck related problems. Drivers (or owners) must see to it that they correct any problems prior to attempting to use the HEVER system. This is necessary in order to guarantee safety. If computers find problems that show up when trucks attempt to enter the system roadway, the truck driver might experience a long delay negatively impacting the ability of the truck to complete a trip as anticipated. As an alternative to not fixing problems caught by the truck's computer or by the HEVER central computer system preventing its use, trucks can use non-HEVER system roads.

The truck's on-board computer system will completely control the truck, virtually eliminating possibilities for accidents while driving on the HEVER system. The truck's computer system will have the ability to communicate with the HEVER system's central computer system; on a cooperative basis, the truck's computer system and the central computer system will manage the safe merging of trucks with other vehicles (collision avoidance).

Design of trucks will include, and hopefully improve on, most of the characteristics of current eighteen wheelers. Safety issues will not allow use of second trailers, due to the automated nature of future truck transport. Trucks will operate as complete units; the cab will completely, inseparably integrate with its freight container. Not only large trucks but also smaller trucks (probably several versions) will use the HEVER system roadway.

If the HEVER system's central computer determines that the traveler or the system must choose an alternate route, it will automatically select alternate routings if the traveler does not or cannot choose an alternate route. If the central system's computer in cooperation with the truck's computer must choose the alternate route and it has more than one choice available, it will attempt to choose the best route. Probably the shortest route and/or the route yielding the shortest time to reach the traveler's destination. The traveler will have been notified of the need to use an

HEVER

alternate route and the truck's computer system will display the choices in a suggested order it calculated. The computer systems will list the available choices in an order as to which appear better to it. As long as an alternate route is available, and a traveler correctly requests the alternate, the HEVER system will execute the request.

Today's economics call for moving as much as possible on one truck in order to reduce the number of truck trips. Due to high weight, this increases the probability of accidents. For example, moving wide loads, such as prefabricated (doublewide) housing units, can prove extremely dangerous. While holding the cost of transporting down for the manufacturer, time and fuel costs to other travelers affected by the moving of doublewides go up.

High traffic volume and congestion increases the probability of the occurrences of accidents. It is very difficult for vehicles to pass these obstructive trucks since they often take up more than one full lane of traffic. The HEVER system will not allow doublewides with their current impractical and unsafe sizes onto its roadway. If you were to transport these doublewides on the HEVER system, manufacturers would need to make them smaller, especially their width.

In order to accommodate the safety of the HEVER system and its users, manufacturers will need to redesign the packaging of prefabricated home subsystems. Assuming appropriate repackaging of prefabricated home subsystems, trucks could transport them under computer control, without manpower that today drives the truck hauling the units and without extra vehicles(s) in back of and/or in front of the doublewide itself. Also the doublewide (probably in pieces) could be transported during the day or at night, as soon as it was ready to be sent. The cost of transporting on the HEVER system will prove lower for making many trips compared to making one unsafe trip on today's Interstate roadways. In order to have the ability to use the HEVER system, manufacturers of prefabricated houses will need to produce them in smaller pieces so that the system can subsequently transport the greater number of pieces to build the same structure as today. This would result in a more efficient, lower overall transportation cost, and effectively the doublewide units could move more quickly.

TRUCKS

Virtually no accidents will happen when using the HEVER system. Without question, the transporting of products and materials such as an unwieldy structure like a doublewide mobile home would, by far, provide greater safety on the future HEVER system than that provided by today's method of doublewide mobile home transport. In all cases, when a truck transporting a doublewide leaves the HEVER system, a truck driver will probably need to drive the truck in order to transport the unit to its final destination. The truck, electrically powered, would be the same truck that was transported on the HEVER system.

When building large enough housing developments, the builder, could have a temporary spur installed for the duration of the developments. This would allow delivery of doublewides and return of empty trucks using the HEVER system. One spur would locate at the manufacturer's factory and another would locate in the proximity of the development. In this case, the units could travel from the factory and the HEVER system could transport them directly to the units' final destination. It would then require handling equipment for the doublewides, only at the location where the housing developer actually sells and finishes the fabrication and building of the homes. Manufacturers could also use the spurs to receive parts and materials automatically at the factory and to receive unmanned or manned empty trucks sent from the housing development.

Truck manufacturers will build complete and self-contained transport vehicles that will have a straightforward simple design. Eventually in the future, separate motors will most likely power individual wheels. All of the powering and braking will come under control of the truck's computer system. Powering more of the truck's wheels individually will result in more reliability and better control. This will allow better handling of the inertial forces while trucks get up to speed when making entry or while slowing down as trucks leave the HEVER system. Also after a truck gets to the required speed while operating off the HEVER system, the truck's computer system can allow alternating use of wheels so motors can take turns resting, so to speak. As a help to the friction type of braking system, the truck's computer system will create drag by reducing power to wheel motors. The slowing or shutting down of all of the motors will, if necessary, lock some or all of the wheels. Under the truck's computer control the truck will have the ability to accomplish improved braking and usually stop more immediately. The wheel system will only be used while operating off the HEVER system or when entering or exiting it. Motors, all-electric, will provide power to all transport vehicles while they don't travel on the HEVER system.

The freight containers must exist as an integral part of any and all trucks. These containers handle the truck's freight. Automobiles will carry a load consisting of people and their various personal items needed while traveling. The total load area for trucks will include the cab area and the integrated back part of the truck used for containing products and materials. The truck and its container must have the capability to leave the HEVER system on its own power and as a single unit. From a standpoint of simplicity all—manufacturers of any automobiles, limousines, trucks, and buses—will build only single unitized and electrically powered vehicles.

Coupled units have enjoyed importance in today's train and trucking industries. Mostly for safety reasons, the HEVER system will not allow trucks with separate cab and trailer on its roadway. As such, companies that own and operate trucks will make great efforts not to allow their trucks to sit idle with empty load containers as much as truck trailers currently sit idle. Electrically powered trucks will have loads as necessary, with or without, driver(s) for transporting on the HEVER system.

The electrically powered trucks will travel, day or night, whenever their owners have them ready and desire to have them travel to a desired destination. Although a driver will need to handle the truck prior to entering and after leaving the HEVER system, trucks will functionally not require a driver except for bringing the truck to a HEVER entrance. After a driver brings the truck to a HEVER system entry or spur entry location personnel can handle the truck while it passes safety checks. At this point, it can enter the HEVER system automatically. Upon leaving the HEVER system, if the driver does not or cannot operate the truck, it will park itself automatically on a parking facility at regular exits or when exiting at a spur. The returning of empty trucks to manufacturers will be less expensive, more convenient, and safer.

TRUCKS

GM has scheduled its 2011 GM Volt automobile to begin production and sales in late 2010. The EPA has given the GM Volt a rating of 230 mpg and it uses a gas generator to produce electricity after the battery has allowed forty miles on a full charge.[27]

Why not apply this technology to a newly designed truck, somewhat like a better built combined GM Volt and an old (improved) design of the GM El Camino? The re-designed GM truck would have a greatly reduced single-speed fixed gear transmission. The truck design might allow at least two versions, a two-door version or a four-door version (crew cab) like that of the GM Colorado. The newly designed truck would have an automobile chassis more like that of the GM El Camino. The truck bed area would allow room for batteries, encased in the truck bed area. The truck bed could have a tonneau cover similar to that commonly found on the bed of a Ford Explorer Sport Trac.

In order to make the GM El Camino light, GM originally designed and built it using an existing automobile chassis design instead of the normal truck chassis design like that for the GM Chevy Colorado and the GM Silverado. Unfortunately, the GM El Camino had a relatively short life as a GM product line. Possibly, it had a short life because it did not have a reputation of "king of the road" like that of the GM Chevy Colorado and the GM Silverado, and because the country was producing and burning very cheap gas. At the time of the GM El Camino, very popular gas hogs traveled the roadways—very popular and probably more profitable to build and market than a GM El Camino.

Consider a GM Colorado truck without its gas powered combustion engine and having a greatly reduced single-speed fixed gear transmission, powered by batteries and an electric motor. With more batteries, driving on battery power hopefully would yield more than the forty miles per charge like that of the Chevy Volt. Also note that miles per charge will vary according to the load on the truck and the driving style of the driver. When the batteries completely run out of the power, the newly designed GM Colorado truck's electric motor would use electricity provided by a gas powered electricity generator like that for the GM Volt. This new

---

27   *"Chevy Volt Gets 230 MPG City EPA Rating," August 11th, 2009, http://gm-volt.com/2009/08/11/chevy-volt-gets-230-mpg-city-epa-rating/#.*

Chevy truck could be designed and built like a two-door GM Colorado or the four-door GM Colorado Crew Cab, which normally has a four-foot bed. A six-foot bed would make the truck quite long. More batteries could reside on the truck bed encased immediately behind the cab. A tonneau cover, similar to that commonly found on the bed of a Ford Explorer Sport Trac, could cover the encased backup gas generator and battery area, and allow considerable room for storage area that could be used as a large trunk.

The GM Colorado is built on a truck bed, but it rides much like typical SUVs that use a similar kind of truck chassis. When originally marketed, the price for this truck design would likely prove higher than that for the traditional GM Colorado, but the truck probably would outperform most of today's hybrid trucks and SUVs. The GM Volt, re-designed and built as a light truck or SUV, would likely give a better performance than an all-electric GM Colorado truck that would have a truck chassis.

Even after trucks are all electric, situations may exist, where the contents transported using the truck have a high enough value so that the situation might require a person onboard for security purposes or for driving the truck when not traveling on the HEVER system. Also, it may prove desirable to have the truck contain a backup power generator capable of generating electricity if needed should the truck's batteries run out of power.

This could help to evolve trucks away from using diesel and gas fuels. What if the truck got an EPA rating anywhere like that for the Chevy Volt? Since the Chevy Colorado now comes with a much larger gas tank capacity than that for the volt, one can only imagine what performance it might enjoy. Currently, it appears that no truck manufacturers have chosen to build any truck having a motor system like that of the GM Volt.

*Like John Henry the steel driving man,
Can I out-drive a computer controlled automobile? Sure I can.*

Chapter 20:
Automobiles

All automobiles will eventually operate as all electric trucks on the HEVER system. The trucks' design must optimize aerodynamics; they especially must adequately handle inertial issues. Automobiles will operate at high speed when on the HEVER system, and at a relatively slower speed when traveling off of the HEVER system. At their top speed, they will operate at a speed somewhere between the current maximum speed of automobiles and that of today's commercial jet airplanes.

Although automobiles have some similarities to trucks, they differ in that automobiles generally carry lighter loads, especially when the automobile has only one or two occupants in the automobile. Automobiles also have an emphasis on transporting people. Trucks contain loads that emphasize the transport of products and materials.

Since humans won't any longer have control of automobiles, accidents caused by humans just won't happen in the situation where automobiles travel on the HEVER system. While traveling on the system and due to computer control, automobiles will enjoy complete safety while attaining much higher speeds, than our current roadways allow. By employing multiple sensors, automobile or roadway functions that can create problems will be monitored and controlled by the central computer system and the automobile computers. Due to advanced computer control, adequate component redundancy, cross sensor monitoring to predict potential problems, and checking automobiles prior to entering the HEVER system, accidents will virtually go away. Attainment of higher speeds will prove more acceptable due to taking control of driving out of the hands of humans and by constraining the automobile on a highly computer controlled elevated roadway while it travels on the HEVER system.

Of utmost importance, automobiles must have extremely advanced design. They will need to transport people at very high speed while on the HEVER system. The automobiles must be small enough, yet carry enough of a load for practical purposes. Weight and height of the automobile have more importance than length. Longer vehicles, such as limos and buses will eventually operate on the HEVER system.

In an article on August 16, 2006, Tesla Motors, Inc., detailed the makeup of the all-electric Tesla Roadster's very unusual battery system. The development of the battery system occurred over a three-year period. Tesla Motors claims that their battery pack has developed as one of the safest, largest, and most technically advanced battery packs. The system uses about 6,800 Lithium Ion batteries exactly like those found in computer laptops. The battery system focuses on safety and durability. If one or a few batteries fail, the battery pack still functions adequately. The Tesla Roadster's microprocessors will identify any non-functioning batteries. Billions of these battery cells are produced annually. Much advancement in batteries has occurred in the past fifteen years and progress will continue to improve rapidly in the future.

Embedded microprocessors control the battery pack and they use a number of sensors to provide monitoring and protection for each and every battery cell. The enclosure of each of the battery cells consists of aluminum instead of the normal plastic, giving each cell greater safety. The entire battery pack, enclosed in steel, results in further added protection from damage. The use of aluminum covering each and battery cell and the steel container enclosing the entire battery facility also serves to operate as a heat sink helping to prolong battery life.[28]

According to an article on September 10, 2008, with its flawed two-speed transmission, an all-electric automobile, the Tesla Roadster, reached an acceleration speed of going from 0 to

---

28   Gene Berdichevsky, et al., "The Tesla Roadster Battery System," August 16, 2006, http://www.teslamotors.com/display_data/TeslaRoadsterBatterySystem.pdf.

60 in 3.9 seconds. It previously had gone a distance of 221 miles. At the time of this article, BorgWarner had signed an agreement to build a more efficient, single-speed transmission that Tesla Motors expects to help provide the Tesla Roadster with 30 percent more power. At the time of the article, Tesla Motors expected to ramp up production to forty per week. Battery powered, the Tesla Roadster, operates very quietly.[29]

An article on March 26, 2009, indicated that Tesla Motors will be producing a Tesla Model-S, seven-seater sedan that will have up to about a three hundred mile range. Production will begin in the third quarter of 2011. It will have a price tag of $49,900 (after a $7,500 tax credit). At the TeslaMotors.com web site, in an undated statement, Tesla Motors indicated the same information, but added that it is twice as efficient as hybrids. The seven-seated Tesla sedan actually handles five adults and jump seats that can accommodate two small children.[30]

In an article on April 13, 2009, Tesla Motors indicated that the Tesla Roadster repeated a 241 mile distance in a Monte Carlo e-rally. It actually improved on its record in that it had enough electric power left in its batteries to go another thirty-six miles, giving it a theoretical distance capability of around 280 miles. If these figures are correct, as Tesla Motors expects, it's all electric Tesla Roadster would soon break its own world record.[31]

Tesla Motors on October 31, 2009 beat their own previous all-electric vehicle distance record of 241 mile range achieved in April, approximately six months after breaking the previous distance world record again. The Tesla Roadster went 313 miles, the distance between Paris and Amsterdam.[32]

---

29   Thomas Ricker, "Tesla readies new transmission, ramping production," September 10, 2008, http://www.engadget.com/2008/09/10/tesla-readies-new-transmission-ramping-production/.
30   Christopher Mascari, "Tesla Model S Sedan Concept: $49,900 Seven-Seater Electric To Hit Streets In 2011," March 26, 2009, http://jalopnik.com/5185844/tesla-model-s-sedan-concept-49900-seven+seater-electric-to-hit-streets-in-2011.
31   Darren Murph, "Tesla's Roadster Rolls 241 Miles on Single Charge, Annoys Petrol Pumps," April 13, 2009, http://www.engadget.com/2009/04/13/teslas-roadster-rolls-241-miles-on-single-charge-annoys-petrol/.
32   Vladislav Savov, "Tesla Roadster Keeps on Rollin', Goes 313 Miles On Single Charge," October 31, 2009, http://www.engadget.com/2009/10/31/tesla-roadster-keeps-on-rollin-goes-313-miles-on-single-charge/.

The above references to these Tesla related articles brings out the fact that technical developments relating to battery operated, all-electric automobile technology occur almost daily. Developments in solar advancement will also occur at a very rapid pace and when practical, will be added to the capabilities of batteries used by today's all electric automobiles.

While most automobiles will eventually operate as all-electric automobiles, the GM Volt offers a look into a reasonably immediate future where vehicles are looking as if they will start to make a dent, reducing the pollution and waste of oil. The Volt, scheduled to start production and sales in late 2010 as a 2011 model, will have the ability to travel up to forty miles on electricity from a single battery charge. If the battery runs out of charged electricity, the Volt will use its onboard gas powered generator to extend its overall range to more than three hundred miles. GM has announced that the GM Volt enjoys an EPA rating of more than 230 mpg for combined city and highway driving, due to the new federal fuel economy formulas. Using the above EPA rating, this will give an electricity cost of 3¢ per mile. GM calls its Volt a game changer. Assuming 30 mpg for automobiles using combustion engines, comparing the electricity cost to gas cost, the electricity cost would equate to a cost of 90¢ per gallon of gas.[33]

In another, newer article about the GM Volt, Bob Lutz, GM Vice-Chairman, gave some insight as to the Volt's expected performance. The GM Volt technology, called a game changer for the automobile industry, will see the delivery of the Volt 2011 model begin in late 2010, less than a year from the publication if this article. Let's hope that the quality of the Volt will live up to the many advertisements that GM has been touting with its latest advertising. The GM Volt would enjoy an even better image if it could employ more batteries. It would increase the weight, but perhaps GM could come out with a version with a longer, more extended chassis. If the Volt had more batteries, probably the charging of added batteries could occur in parallel by banks, thereby not increasing the expected time to charge the batteries fully. Wouldn't it be great if the Volt could travel, without using its gas reservoir,

---

33   *"Chevy Volt Gets 230 MPG City EPA Rating," August 11th, 2009, http://gm-volt.com/2009/08/11/chevy-volt-gets-230-mpg-city-epa-rating/#*

eighty or 120 miles in the electric-only mode before needing a recharge? At any length, this technology is exactly what is needed now. Even though it will help us evolve toward all-electric automobiles, like hybrids, they will continue to use some gas. While helping our country wean itself off of gas, using any gas makes the technology a temporary stopgap. The more the Volt runs on electricity without needing to use its gas generator the more it will help to use less gas, conserve the world's oil resources and reduce our country's pollution.[34]

The references to the immediate previously mentioned articles show how new developments occur almost daily. Assuming that GM can produce the Volt with an mpg performance that meets expectations, design considerations will give way to quality issues, which will then loom as a very significant issue. Expect initially high prices; but as the Volt moves into mass production, prices should eventually come down. While employment of this technology will not replace the need to provide the production of all-electric automobiles, its design and development will allow this move away from gas usage to occur in an evolutionary rather than in an abrupt manner. We will pollute less by using less gas and will use more, lower cost electricity. This should also encourage the oil and gas industry to move in an evolutionary manner to other uses for oil.

An article "GM Works to Make Some Noise" appeared on November 25, 2009 in the *St. Louis Post-Dispatch* indicating that the largest advocacy for the vision-impaired will want noise added to the too-quiet automobiles like the GM Volt; apparently blind persons have had near accidents due to not knowing that these quiet vehicles are approaching. Shouldn't motorist always control their vehicles and bend over backwards to give pedestrians the right of way? What happened to the push toward defensive driving? Adding noise seems obnoxious, but it will probably happen. One wonders why drivers can't drive more carefully and not assume a person can see them. This phenomenon of inconsiderate driving can happen in many places; on shopping center parking

---

34   "Bob Lutz Implies Chevy Volt Will Get Between 40 and 50 MPG in Charge Sustaining mode," October 27th, **2009**, http://gm-volt.com/2009/10/27/bob-lutz-implies-chevy-volt-will-get-between-40-and-50-mpg-in-charge-sustaining-mode/.

lots drivers appear to believe that they have the right of way and that pedestrians have the responsibility to get out of the way.[35]

None of the currently proposed automotive industry changes gets us to high speed nor do they, in any significant way, attack the safety problem. It would be nice to eventually have the automotive industry producing more highly computer controlled all-electric automobiles. However, essentially as now designed, plus an addition of some changes for use as a prototype for testing, perhaps an upgraded GM Volt could eventually operate as the first automobile to gain full implementation on the HEVER system.

Automobiles will have space for clothes and other personal necessities and optionally they will have telephone, fax, small office facility, television, radio, computer entertainment, and a lavatory. Its main interior will have seats like that of today's automobiles. Hopefully automobiles will have many improvements. Automobiles will have the possibility for many new options due to highly available low-cost electricity. In the future, more standardized automobiles will have a main unit length and width like today's automobiles.

The automobile's computer system will automatically check for problems and will communicate with the central system to see if its computer system detected any problems. The central system's computer will from time to time need to update information in the automobile's computer database. Automobiles will not be allowed onto the HEVER system if their computers and/or the HEVER system central computer system detect any problems. The problems must be corrected before any automobiles can travel on the HEVER system. This rule will help to guarantee safety. If the computer systems detect some problems, a long delay may result, negatively impacting the ability of the traveler's automobile to complete a trip within the expected time frame. When drivers attempt to have their automobiles enter the system roadway, as an alternative to not fixing problems caught by the automobile's computer or the HEVER system central computer, automobiles can use roads existing off of the system roadway.

---

35   Chris Woodyard, "GM Works to Make Some Noise," USA TODAY, November 25, 2009.

AUTOMOBILES

The automobile's computer system can automatically or, in some cases, a traveler can manually, select and cause use of alternate routings. Automatic alternative route selection will happen when the computer systems determine that it is necessary to take such an action. Manual alternative route selection will occur when travelers determine that they need to change their destination or exit because of a previously unplanned emergency or perhaps they want to eat at a restaurant. When requesting the use of manual selections of destination changes, the traveler must make certain of the validity and ability to do the requested change or it will not occur.

The automobile's computer system can communicate with the HEVER system's central computer system. On a cooperative basis, the automobile's computer system will allow the central computer system to manage the safe merging with other automobiles, buses, or trucks (collision avoidance). The automobile's computer system will completely control the automobile virtually eliminating possibilities for accidents.

*I don't have an automobile or truck. Why all the fuss?
No problem, let's take a bus.*

Chapter 21:
Buses (and Limos)

The design of buses and limousines will require high safety standards. Buses and limousines, radically different today, will function almost in an identical way on the HEVER system.

Bus operation on the HEVER system will most likely happen after truck deployment and before or after adding automobile traffic. After proving the viability of bus operation, its computers and the central computer system will control and operate buses as part of the HEVER system. Depending on the sequence of adding most additional types of vehicles, developers will have implemented bus traffic. Original implementation of bus traffic will prove easier for relatively short distances. When implementing addition of bus traffic standard entrances to and exits from the HEVER system must exist with a capability to accommodate these buses.

Over the long term, buses will normally travel short or long distances on the HEVER system. High speed bus transport between the suburbs and the inner cities (more stops) will prove most beneficial, but this will take longer to implement than the longer distance bus capability. Allowing automobile traffic onto the HEVER system will likely start before addition of buses and after trucks; this will allow time for considerable simulation, experience, and statistics gathering from the freight system (and possibly a bus test vehicle). Implementing electric truck and electric bus service on the HEVER system will cause significantly lower oil waste and its subsequent pollution effect.

HEVER

Implementation of the operation of trucks on the HEVER system will give way to lower costs. Considering all of the economic gains for trucks, its total annual monetary saving should far out distance the total annual dollar savings of that for buses and automobiles combined. Just because of its life saving capability, automobiles (and possibly buses) operating at high speeds on the HEVER system will, by far, bring more personal benefits (especially the elimination of fatal accidents) than the combination of the freight and bus additions. Implementation of the bus addition to the HEVER system will prove useful to persons not able or not wanting to drive. Maybe they do not have the desire to own an automobile or possibly they cannot afford an automobile. Also, in general, people currently prefer to travel in the privacy of their own automobiles. In the long run, this need heavily perceived by travelers will continue to make the automobile the preferred, main mode of transport for most passenger traffic on the HEVER system.

Individual automobiles will transport travelers primarily because they will want a vehicle that is clean (or at least it contains their own dirt) and they will generally prefer not to travel with strangers. Also, travelers usually will want to take baggage with them and they dislike the need to check and retrieve their baggage. They will often want to take some other personal items that they think they will need. Travelers dislike using buses, trains, or airplanes because sometimes their personal belongings get sent on a later bus or even end up at a wrong location. People must then experience the nuisance of waiting for their (sometimes lost) baggage. In the long term, future automobile type vehicles will make provisions designed along the functional needs of today's business and pleasure travelers.

Similar to truck and airplane travel, taking a bus even on the HEVER system, will also waste time due to the necessity of making intermediate stops at bus stations. Buses generally operate more economically when they can make stops at more places where people can enter and leave. Too many stops can make bus travel less desirable. With regards to capacity, buses will normally handle a larger number of passengers than individual automo-

BUSES (AND LIMOS)

biles and trucks. Sometimes buses can offer an option as a better alternative to trains, especially on the future HEVER system.

Buses will more safely transport people on a raised roadway that will use the highly controlling HEVER computer system.

Of great importance, developers will phase in the types of vehicles according to the type of use and amount of traffic volume. Those vehicles that will more quickly grow volume can have higher priority for appearing earlier on the HEVER system. The requirement to transport passenger traffic on the HEVER system during what is now rush hour traffic will be handled by the HEVER system design after truck traffic implementation for commercial traffic. Automobiles will most likely implement after trucks and maybe even after buses. Because of employment of electromagnetic braking and switching, trucks, automobiles, and buses will safely and efficiently travel at high speed on the HEVER system.

Since trains are constrained on rails, they generally are safer than buses. However, this situation will reverse upon HEVER system implementation. Buses will prove safer than trains.

Today, buses can make some situations more dangerous for smaller automobiles and motorcycles. Currently, other vehicles can find it difficult to pass slow moving buses. When buses operate abnormally slow, it can cause other traffic to slow down. When they operate at too high of a speed, they can create a dangerous situation. These issues will not exist after HEVER implementation.

Currently, depending on schedules or chartering, buses can operate at any time of night or day. In the future, a computer controlled system of elevated roadways will allow buses to operate on or off of such a system. On the new system, buses will have the ability to make intermediate stops, or if desired, they will have the ability to operate non-stop from start to finish of a journey.

Delivery of buses from the manufacturer can be highly automated when delivered using the HEVER system.

The design of buses will require high safety standards.

In the future, buses inertial design requirements will be greater than for automobiles and have similar design requirements to that for trucks.

HEVER

Bus operation on the HEVER system will happen; future studies will first need to prove the feasibility and practicality. Then buses will go through rigorous testing. After proving the viability of bus operation, the central computer system and bus computers will control and operate buses as part of the HEVER system. Developers can implement bus traffic after implementation of truck and automobile traffic. Easier to originally implement bus traffic for relatively short distances, standard entrances to and exits from the HEVER system must exist to accommodate these buses.

Over the long term, buses will normally travel short or long distances on the HEVER system. High speed bus transport between the suburbs and the inner cities (more stops) will prove most beneficial, but this will take longer to implement than the longer distance bus capability. If possible and practical, addition of automobile traffic to the HEVER system will start before buses and after trucks, allowing considerable simulation, experience, and statistics gathering from the freight system (and possibly a bus test vehicle). Implementing electric truck and electric bus service on the HEVER system will cause significantly lower oil waste and its subsequent pollution effect.

Why acknowledge design differences, often called incompatibility?
About half of marriages last. Do people compromise with agility?

Chapter 22:
Automobile and Truck Non-compatibility

Automobiles and trucks in traffic do not exhibit complete compatibility with each other on the same roadways, and as noted earlier, the first justification for sophistication of roadways was actually for commerce not for passenger traffic. The first HEVER system implementation will happen for truck traffic. Implementers will add bus type traffic (without humans aboard) somewhat early in the testing on the same test roadway. Later after proving the safety for truck traffic, developers will possibly add bus traffic. Following this first testing implementation or possibly parallel to the completion of the testing of truck traffic, a longer part of the HEVER system might have the ability to have vehicle(s) accomplish successful operation in several places. At some point, automobile traffic will start into its testing phase on the HEVER system roadway first used for trucks and possibly buses. Truck and bus vehicles on the HEVER system typically need to originally experience higher traffic volumes and show an ability to carry much heavier loads than those for the automobile traffic on the HEVER system. Also, it will eventually turn out that more than one HEVER system lane will accommodate many types of vehicle traffic. An additional lane(s) will allow a higher traffic volume and will provide a much greater ability for doing construction and maintenance. Initially building two lanes in each direction will be more practical even if in the initial HEVER system roadway implementation would only need one lane in each direction.

Normally trucks and buses will use the same lane(s) as automobiles. Since trucks, buses, and automobiles will travel at the

HEVER

same speed on the same lane(s), headway distance, in the front and back of trucks and buses will usually need to have a longer length than that for automobiles. Required headway distance between vehicles may need to vary depending on the actual weight of the vehicle and the truck, bus, or automobile contents.

Hopefully all vehicles will have the ability to operate at the highest allowable speed; this would allow the most efficient use of the HEVER system roadway. Design of trucks and buses might be critical for determining if both a fast and relatively slow lane must exist. If the different HEVER system vehicle types (truck, bus, and automobile) need to operate at different speeds, developers will need to build both a fast lane and a slow lane. This slow lane will also handle all types of traffic and will serve as a slow down lane for vehicles preparing to exit. For instance, possibly trucks and buses may need to travel at 200 mph to 300 mph while automobiles will travel at 400 mph. Trucks and buses, while they can probably operate at higher speeds than they do on today's highways, on a future HEVER system they may need to, because of size and weight, operate at lower speeds than automobiles while on the HEVER system built to handle automobiles, trucks, and buses. While trucks and possibly buses will likely require a longer, wider, and higher design than that for automobiles, they all should not cause the need to have lane(s) with differing widths; the width required for the widest allowed vehicle should determine the maximum required width. It will likely dictate a single standard width for all lanes.

Inertial forces are different in these HEVER system types of vehicles, causing a need for different size and strength standards in the HEVER system design. Trucks, completely self-contained vehicles, will require a cab or front space for a driver required to drive the truck after the truck exits or before it enters the HEVER system. While on the HEVER system, trucks and buses will operate with or without drivers. None of the persons present in trucks, buses, or automobiles will control the vehicles while traveling on the HEVER system. Unless some emergency occurs while traveling on the HEVER system, all vehicles will travel under complete control of the HEVER system.

AUTOMOBILE AND TRUCK NON-COMPATIBILITY

On the HEVER system, automobiles will inherently come in smaller sizes than that for trucks and buses. Designers of automobiles (and buses) will need to provide as low, narrow, and as light of a design as possible. This will lead to designs, in which passengers fit into vehicles much like today as long as manufacturers build safe, standard sized vehicles. It might also be possible to design somewhat large limousines (standard size) as buses, since trucks will require a somewhat different design.

*For baggage, any vehicle should have storage that is big.
This huge SUV has lots of storage space, what a rig.*

Chapter 23:
Operating the HEVER system

The first passenger all-electric vehicles (preferably small automobiles) designed for traveling on the HEVER system will allow for some baggage and as much convenience as possible and practical. Initial usage of the roadway would then enjoy sooner and, therefore, greater economic advantages. This new type of travel would also satisfy a greater range of commuter requirements; people could commute at much greater distances than in the past. This would not preclude the eventual addition of larger personal vehicles for transporting families between cities.

In the future, as the HEVER system matures and becomes the preferred high speed method of transport, the situation of refueling will prove less and less of a problem. While the vehicles are on the HEVER system, vehicles will not require use of its batteries; the system will actually charge the vehicle's batteries. The vehicle owner must have an account with the HEVER system's central computer system for this; automatic fee charging will then occur, charged to the owner's account. Without presence of an owner in the vehicle, a non-owning traveler in the vehicle must present some form of ownership identification (the owner may have provided a traveler with his personal ownership identification). As an alternative the non-owning traveler could use a method of payment such as a debit or credit card to pay for the fee. The central computer system will check to see that it can easily charge for the entire trip. Refueling and paying for fuel is today a necessary nuisance. It costs perhaps ten to fifteen minutes, more or less, and is virtually a complete waste of time in that not much else other than refueling can be done during that time.

In the future, using electric vehicles, refueling can happen at kiosk manned or unmanned stations while driving on today's streets (while not on the HEVER system). The kiosk will eventually have the ability to replace automobile and truck batteries in less time than for today's gasoline driven automobiles. This will only occur when vehicles can have replaceable batteries. At the kiosk, upon opening the battery compartment, the batteries needing replacement would pop up indicating that the batteries need replacement because they either need to have an electrical charge or the batteries are bad. In either case, replacement batteries will do the job. This could turn into a normal process even before people use their vehicles on the HEVER system. Travelers could pay for a replacement(s) by using cash or a credit card at a kiosk if traveling off of the HEVER system or by using a credit card or an owner system Id account number if the owner, at the time, travels on the system. If the traveler(s) must exit to charge an out of charge battery or to replace a bad one, cash (if kiosk is manned), debit, or credit card charges could be handled at the kiosk for the vehicle that exited the HEVER system. It will be feasible to place self-service kiosks at many, many locations, for example at today's gas stations.

Normally charging of batteries will occur while vehicles travel the HEVER system and the batteries will receive as full a charge as possible. Eventually high availability of battery replacement in some standard form will happen at most system exits. This will only happen after vehicles and batteries are designed to use easily replaceable batteries.

Brakes? Who worries about having a good set of brakes?
Farmers don't; they mostly worry about their tractor's rakes?

Chapter 24:
Driving On and Off the HEVER System

While persons are driving automobiles, buses, and trucks on roadways other than the HEVER system, people will sometimes likely want to stop at eating places along the way. This need will exist for the foreseeable future. Even after the HEVER system becomes an institution, people will still sometimes want to use the existing Interstate system. The gas stations on the Interstate could evolve into facilities that provide food, oil, gas, restaurants, battery replacement, and other sundries. If travelers wish to use these facilities they will need to let the system know, so that they can have their vehicles appropriately exit the HEVER system.

The Interstate system needs to stay in place for the foreseeable future. It can provide backup for the HEVER system and will likely find use for slower and short distance driving. Perhaps officials can even raise driving speed limits on the Interstate system when electric vehicles become commonplace. Electric vehicle computer systems will have more safety devices. Electricity will be more available, cheaper, and cleaner than gas used by vehicles. Individual computer and sensor components will cost pennies per unit to manufacture in high volume. They will check the viability of vehicles entering the HEVER system.

In the future, electric motors will probably more individually operate each of the vehicle's wheels. Perhaps if individual motors powered individual wheels, the motors might give vehicles more efficient and more reliable service. Motorizing each individual wheel would allow the vehicle to operate, in some cases, even if three wheel systems failed. The vehicle could sometimes still operate as long as one wheel still operated. This process would only occur while people operate the vehicle off the HEVER system. A considerably simpler single gear transmission will allow greater efficiency and reliability for all fully electric automobiles.

Current, existing conventional friction based brake systems used in cars have significantly improved through the years for today's vehicles. Hopefully, in the future, designers will provide even more improved brake systems. These improved brake systems will need to happen. Our society will need to have adequate, safe brake systems for these future individualized vehicles on any mass transport systems. We will still need friction brake systems when driving vehicles off a high speed mass transport roadway like the HEVER system. A friction brake system might have the ability to actually serve as an emergency brake system while vehicles are on the HEVER system.

The source of power for the electromagnetic brake systems will normally come from the HEVER system. If the HEVER central system loses power, the vehicle battery power will power the brake system. In an emergency, the friction based braking system then might serve as a backup braking system while vehicles are on the HEVER system.

Use of the electromagnetic brake system will provide a better, more reliable braking system than any of the current friction types of brake systems. Some friction type of brake system will need to exist on vehicles for use when the vehicle operates off of HEVER system. A non-friction brake system will employ electromagnetic propulsion basically operating in reverse for braking devices. This braking system will exist on the vehicle, on the HEVER system roadway, or, more likely, on both. Vehicles will need to make use of battery-powered motor(s) and a friction brake system for use while vehicles are operating off the HEVER system. If the main HEVER system central control system fails, the electromagnetic based braking system on the vehicle (possibly along with the friction based braking system) will operate to safely reduce the speed of vehicles or bring the vehicles to a complete halt. If the main computer system was still operational and the vehicle's computer system failed and/or the vehicle's electromagnetic brake system

DRIVING ON AND OFF THE HEVER SYSTEM

failed, the main computer system might have the ability to take control of the HEVER system's friction based electromagnetic brake system and/or the vehicle's electromagnetic brake system. Together or individually, they could control vehicle movement, speed, and stoppage of the vehicle.

Electromagnetic brakes have additional benefits over friction brakes, in that they will easily function normally on roadway surfaces whether oily, wet, or in some other way slippery. Using friction based brakes often won't easily handle these conditions. The electromagnetic brake system will still operate successfully even when under snow and ice conditions.

Basically, electromagnetic braking systems will operate by causing the electromagnetic propulsion system to operate in reverse. This will allow the system to slow or stop vehicles, as needed, while they operate on the HEVER system.

Special emergency exits will exist for the purpose of letting vehicles exit to a roadway (probably the Interstate). If power fails, the computer system at exits could allow the vehicles, using battery power, to automatically use the nearest exit. Deployment of emergency exit use can happen any time a system power outage occurs, especially when trying to operate without central system power for an extended period of time. Signals to activate automatic exiting could happen by wire or wireless transmission to vehicles that they should use emergency or normal exits. If an emergency occurs, such as loss of HEVER central system electricity, all vehicles would need to stop.

When all vehicles computer systems can recognize the reality that an emergency situation has occurred, all vehicles can safely and automatically exit. Drivers normally will have the ability to drive the vehicle after leaving the HEVER system roadway. Emergency parking will be available as needed. In the emergency where a parking lot fills up, vehicles' drivers or kiosk attendants will need to handle vehicles manually or vehicles will continue to the next exit.

Unlike today's travel in an automobile and like today's travel on a train, road signs for information, stop signs, and traffic lights won't need to exist for travelers in vehicles on the

HEVER

HEVER system. Computers containing the location information can display on monitors inside the vehicle and these computers can provide information such as current location, estimated remaining time needed to reach a destination, destination arrival time, potential alternate routing, potential alternate exits, emergency exits, emergency procedures, entertainment, and work. Travelers can instantly access travel and other information, accessible from the HEVER system central control system's and the vehicle's computer systems. Displayed on CRT monitors in the travelers' vehicles, the latest information will appear on the CRT's display at the traveler's command. Up to date information provided by the central and vehicle computers that control the HEVER system will make stop signs and traffic signals virtually unnecessary while traveling on the HEVER system.

*Manufacturers want to send their products safely, fast, and at low cost.
At their customers' stores, they hope their products won't get tossed.*

Chapter 25:
Manufacture, Distribute - The Kings

Development of the HEVER system for trucks will originally emphasize manufacturing and distribution interests. This will significantly help to reduce truck traffic volume on the Interstate system and more efficiently automate transport and delivery of freight. Somewhat immediately, it will help to reduce pollution and will reduce the burning of precious fossil fuel that we need to save for other uses (for example lubricants and plastics). Also, it will reduce the number and severity of accidents across the nation. In many cases, it will be possible and practical for trucks to traverse the HEVER system without anyone being on board the trucks.

They will operate non-stop from origin to destination. The HEVER central system computers and the truck's computers will completely control and effectively transport them as freight. As an option, if owners (probably companies) desire to have the HEVER system transport its trucks without drivers, drivers will still need to drive the trucks before they enter and after they leave the HEVER system.

Manufacturers, distributors, and providers of materials or parts will have, as an option, the ability to use spurs into and out of major manufacturing and distribution facilities, shopping centers, and industrial warehouses. Developers will construct these spurs as normal components of the HEVER system. Except for some emergency exits, all spurs must operate as manned facilities and must have kiosk stations with computer systems that have the ability to accomplish vehicle checking. These locations must also provide temporary and automated parking facilities.

HEVER

Implementation of the operation of trucks on the HEVER system will yield lower costs to truck owners and their customers. Its total annual monetary saving will likely far outdistance the total annual dollar savings of that for the operation of buses and automobiles combined. Just because of its life saving capability, the automobile operating at high speeds on the HEVER system will, by far, bring more personal benefits than the combination of the freight and bus additions.

Implementation of the bus addition to the HEVER system will prove useful to persons not able or not wanting to drive. Maybe they don't have the desire to own an automobile or maybe they cannot afford an automobile. Also, in general, people currently prefer to travel in the privacy of their own automobile. In the long run, this current privacy need, heavily perceived by many travelers, will make the automobile continue as the preferred, main mode of transport for most passenger traffic on the HEVER system. In the long term, future automobile type vehicles will make provisions designed along the functional needs of today's business and pleasure travelers.

Of great importance, developers will phase in the types of vehicles according to the type of use and amount of traffic volume. Those vehicles that will more quickly grow volume will likely have higher priority for appearing earlier on the HEVER system. The requirement to transport passenger traffic on the HEVER system during what is now rush hour traffic will be handled by the HEVER system design after truck traffic implementation for commercial traffic. Partly because of use of electromagnetic braking and switching, automobiles will safely and efficiently travel at high speed on the HEVER system.

What good is simulation? What does it do?
Often, when designing large systems without simulation—pooh-pooh.

Chapter 26:
Simulation

The HEVER system, automated through central computer control, will allow vehicles to operate under their own computer control in cooperation with the HEVER system central computer control. After implementation of the HEVER system, simulation models established to aid in the initial design of the HEVER system, will be adjusted (improved) to allow use of actual pertinent data gathered by the HEVER system central computer system.

The design of the HEVER system must learn from the past that designers must consider future population growth in order to allow good extensibility. Design and implementation of the HEVER system will use a phased approach. In the long range, system design and development must allow creation of a highly engineered overall system design.

Prior to implementation, the designers will do as much simulation as time and economics will allow. After completing the setting up of a working model of a portion of the overall system, simulation studies will continue even after HEVER system implementation. Growth (size) of the simulation model will proceed on an ongoing basis. The electrical nature of the vehicles and the HEVER system will lend itself favorably to gathering statistics and using simulation techniques to study and improve the HEVER system.

Truck, automobile, and bus traffic detail activity values provided as components (statistics) will be needed by the simulation model. This will allow continuous study of each type of vehicle and their interactions and to study small to large aggregations of traffic components. The simulation model will provide much insight into the design and development of software to handle inertial and collision avoidance issues.

After the development of the HEVER system has reached the level allowing travel between two or more cities, the simulation model will have grown to a large system. In order to control the accuracy of the model management of the model, components will likely have to consist of a network of sub-models integrated into an overall model. Simulation studies will continue while and after building a full roadway system.

Breaking the simulation into sub systems will help determine an approximate number of required computer systems, give indication of required amount of computer power, and suggest where to locate central or regional computers. It may even determine that a central controlling computer control system is unnecessary.

Gathering of data for use by a simulation system will happen as a natural byproduct of the HEVER system because computer data gathering will occur naturally as vehicles travel on the electrically powered system.

Management must hire a top number of mathematicians and engineers in order to support required disciplines such as simulation. Development of a simulation model will require a depth of knowledge and experience in mathematics, statistics, and simulation. These design mathematicians, statisticians and engineers will design and implement the HEVER simulation models that will help to guide actual system design.

*We need new sources of energy; let's go solar.
Among other benefits maybe we'll save the bears we call polar.*

Chapter 27:
Solar Energy

The technology of solar derived energy is advancing very rapidly. Several articles and press releases will show some of its progress. The following articles are intended to show how Companies like IBM are developing solar technology. In the future, NASA is intending to make use of advanced solar devices. Solar energy, which will continue to improve, will likely find usage by the HEVER system which will need low cost electricity.

The HEVER system will to some point employ electric utilities to provide electricity service. However, engineers will plan and build some or all of the power generation facilities needed to operate the HEVER system itself in order to better guarantee available power. The electric utility industry and HEVER system electric generating facilities can then work together to provide each other some modicum of backup. Wind and solar collectors will be built along the HEVER system, giving free energy to the system. The system can produce virtually pollution free electricity energy. Wind and solar collectors on the HEVER system might collect enough energy to operate the system completely and to provide some unused electric power. Current solar energy collection devices, by themselves will not likely provide enough electricity to operate the HEVER system fully. Generators using reproducible resources could provide some power. At this time, it is unclear whether or not hydrogen fuel cells can serve to generate some of the needed electricity, but if practical and non-polluting, hydrogen fuel plants could locate at various places along the HEVER system to produce electricity.

HEVER

Vehicles traveling on the HEVER system should eventually provide as much power as possible, perhaps by having solar collectors on the vehicles. It may prove possible to have solar collectors on the sides of the roadway. While traveling off the HEVER system, electric vehicles will travel farther on a given set of charged batteries if solar collectors could exist on the vehicles to help provide energy. In addition to the above potential energy production, low polluting or nonpolluting replaceable fuels could generate electric power at strategic locations along the HEVER system. Unfulfilled energy requirements may cause the HEVER system to obtain electricity from the electric utilities. Virtually all of the energy for operating the vehicles while traveling on the HEVER system might come from the HEVER system itself, eliminating most of the need to obtain electricity from the electric utilities. If so, the HEVER system will charge vehicles' batteries as fully as possible while they continue to travel on the system. To the degree that the system can add some electric charge to vehicles' batteries, travelers' vehicles will have the ability to go greater distances after they leave the HEVER system.

If needed by the HEVER system roadway surface, when temperatures drop below freezing, the HEVER system might provide heat tracing. Developers might need to implement heat tracing to keep the roadway or any of its components from getting covered by ice and snow during winter. The heat tracing design might allow automatic computer control or, more likely, maintenance personnel will decide when to turn on the heat tracing feature of the HEVER system. In most locations, it would only need to operate during the winter. Also, winter usually will create an extra high demand for electric usage on the electric utilities. If some of the energy generated by the HEVER system is not needed, the electric utilities can use it at little or no cost.

Much energy research and development is ongoing by many companies and government agencies. IBM currently is working on solar research, much of which still has to get out of the lab, and into production. Quite often, IBM doesn't actually produce the things that it invents (patents). Rather the company sells licenses to other companies giving them the right to produce the actual

SOLAR ENERGY

products. IBM extensively researches processes and methods, which it believes it will soon make available for production in the near future. IBM intends that this research will vastly improve our ability to harness the power of the sun. In the referenced article, IBM alleges that it has made a research breakthrough in reducing the cost to harness electric energy from the sun. IBM indicated that its new methods will reduce, by a factor of ten, the number of required photovoltaic cells and other components.[36]

Solar energy is produced from the sun's rays generally using a silicon collector cell consisting mostly of silicon with some embedded metals. As solar collectors continue to improve, companies, in solar collector industries, will produce increasingly improved solar collectors. In the coming years, these will contribute significantly to availability of these improved solar collector devices.

NASA has designed an all solar powered airplane prototype that it intends to use for distant planet exploration. Plans are in the works to have at least one and probably several more exploring the planet Venus, which has far more sunlight than the Earth. Since Venus has much more sunlight than the Earth, such an airplane can cover a wide range of Venus and operate virtually for the foreseeable future. While flying at high altitudes, the airplane can obtain phenomenal amounts of energy from Venus' sun.

The airplane, designed at the NASA Glenn Research Center, uses sunlight to produce electric energy to power the airplane. It will have the ability to fly continuously. The airplane could run out of energy if it operates too low to make extended terrain studies for too long of a time period. Some issues, like flying too low, will require considerable planning to keep the airplane flying.

Venus has situations that remain unknown to our scientists, both on its surface and in its atmosphere. With appropriate measuring, the airplane could investigate and figure out if Venus, long ago, had water. The airplane might also determine if Venus had volcanoes in its distant past.

---

36   "IBM muscles into CIGS solar-cell market," May 15, 2008, http://www.physorg.com/news130086323.html

The tops of clouds move much, much faster than the bottoms do, near the surface of Venus. NASA does not completely understand this somewhat strange phenomenon. The airplane's instruments could measure some of these atmospheric unknowns and allow our scientists to gain a better understanding of them.

The aircraft will make great use of radar to investigate very far away Venus surfaces. This would allow the airplane the ability to view many surface areas since it will be able to orbit rather than operate in a geosynchronous manner. Venus exploration will not have a restriction of only looking at one area. This type of airplane can make unusual progress in accomplishing much better exploration. Comparing Venus' past with that of the earth, Venus' has many similarities and also many differences. Exploration of Venus by this plane can help scientists to understand better the past development of atmosphere and weather conditions on Earth. If the first solar powered airplane proves successful, many more airplanes could investigate Venus and do joint investigations.[37]

---

37   *"Solar Airplane Developed for Venus Exploration,"* Dr. Geoffrey A. Landis, NASA Glenn Research Center, January 21, 2005.

*Electromagnetics as a concept has been around for so long.
If we use it to power trucks and automobiles, we can't go wrong.*

Chapter 28:
Electromagnetics

We can use one of the most important fields of technology. No, we must use this technology, to fully and physically control land based vehicles such as trucks, buses, and automobiles. This technology will propel these vehicles forward to speeds of 400 mph or even more and use computer control to virtually instantaneously slow any number of vehicles down or bring any or all of them to a virtually instantaneous halt.

According to ancient legends, about four thousand years ago the Greeks made magnets and eventually compasses from magnetized metal found naturally in an ore. The magnetic compasses found great use in navigation and helped to make magnets themselves commonplace. The Greeks called this unusual magnetic ore magnetite. In ancient times, some cultures made great use of magnets while some people even thought magnets possessed evil powers.

Accidently in 1820, Hans Christain Oersted discovered Electromagnetics, the technology that will do the job of transporting people and material safely and at high speed. Scientists originally believed that electricity and magnetics had nothing in common with each other, and enjoyed a totally unrelated phenomenon. Virtually everyone believed that electricity only included such things as lightning bolts and powering motors and light bulbs. They also believed that magnetic metals did not have electrical properties, but that they had non-electrical, almost magical, properties. People found that they could use them in compasses and for other magnetic related purposes. The world now understands that electricity relates closely to magnetism. When using a process of properly passing an electrical charge through certain metals, the process adds a new characteristic (property) to the metal causing it to become a magnet. We call this phenomenon, in which electricity causes magnetic fields to occur in certain metals, Electromagnetics. The following excerpt comes from an article in which Dr. Frederick Gregory discusses Hans Christain Oersted's accidental discovery of Electromagnetics in 1820.

> The actual discovery of electromagnetism was made during a lecture demonstration that Oersted was conducting for advanced students during the spring of 1820. It is perhaps the only case known in the history of science when a major scientific discovery was made in front of a classroom of students.
>
> Oersted's first inclination was to characterize the force affecting the needle as an attraction of some sort. But he found that moving the wire to the left or right, all the while keeping the wire parallel to the needle's original position, did not affect the nature of the deflection. Hence, the force could not be an attraction between one pole and the wire, for in that event the attracted pole should follow the wire.
>
> The needle did swing in the opposite direction in two situations: 1) when the wire was positioned beneath the needle, or 2) when the current was reversed in the wire. From these and other manipulations of the apparatus, Oersted announced his conclusion regarding the force:
>
> What is it that Oersted had discovered? Through persistent and repeated efforts subsequent to the classroom experience Oersted clarified the precise nature of the effect a wire conducting electricity had on a magnetic compass. He found that a wire carrying an electric current affected a magnetic needle located below the wire by causing it to swerve to a position perpendicular to the wire.[38]

---

[38] Dr. Frederick Gregory, "Oersted and the Discovery of Electromagnetism," University of Florida, www.clas.ufl.edu/users/fgregory/oersted.htm.

ELECTROMAGNETICS

Electromagnetic propulsion will use electricity and high-powered magnets to power the HEVER system. It will allow use of its roadway by electric vehicles. While not likely, possibly vehicles other than electric vehicles could use the new roadway. Vehicles do not provide their own power while they use the roadway. Non-electric vehicles would need to have all of the components necessary to allow levitation and computer monitoring and control; this would not likely happen. Probably more likely, design and development of skeleton vehicles, so to speak, could provide transport of non-electric vehicles. These skeleton vehicles would need to have all of the necessary computer hardware and software and the electromagnetic devices with full capability to operate on the HEVER system properly.

The HEVER system will transport vehicles not as an operating vehicle, but rather as freight that the electromagnetic propulsion uses to push the vehicles along the automated roadway. During normal operation, the system does not even need to use the vehicles' wheels at all. The HEVER system will use electromagnetic propulsion that will transport the vehicles in a process conceptually similar to the propulsion system used by maglev trains. The roadway surface will consist of a metallic surface above which the automobile will levitate using an electromagnetic form of propulsion.

Using electromagnetic propulsion to move vehicles forward on the roadway, the HEVER system will provide electric power for operating the electromagnetic devices and will ensure, to the degree needed, that vehicles' batteries are charged as much as possible for off system travel.

A well known electromagnetic technology called maglev (magnetic levitation) propelling high speed trains, allows them to exceed speeds of 400 mph. These trains can move so fast because they employ electromagnetics to not only propel them forward but also to make them slow down or stop or even operate in reverse. Hence, we will have a non-friction braking system. The maglev technology allows these trains to go slightly (but highly controlled) airborne eliminating ground friction resistance and wear. That is, the maglev trains glide, as much as a half inch or, in some cases, almost as high as four inches above the roadway surface depending on the electromagnetic propulsion method and associated devices used. Keep in mind that a train is much heavier than an automobile or even a truck. As such, it should be easier to levitate and propel an automobile or truck by using electromagnetic propulsion. Scientists have named this lifting above the ground phenomenon "levitation."

Now comes some bad news. Maglev trains have all of the old disadvantages of old conventional trains except for the fact that they can go faster. Unless the maglev trains are express nonstop trains, they still must stop at many stations. They do not have the characteristics of freedom virtually to go anywhere, like cars, trucks, and buses that have built in flexibility of directly reaching almost limitless locations. People and materials ready for transporting still have to make use of non-train forms of transportation to transport freight and people to train stations. This difficulty of getting the people and materials to a train station prior to starting the trip, does not happen when using the HEVER system. At the end of a train trip, travelers and materials must use yet another, separate vehicle that will then transport the people and materials from the station to a final destination. Very similar to that for airplanes, trains must load and unload cargo and people at stations along the way. Sometimes trains can use spurs, which manufacturers and distributers use as starting and ending locations for products and material. The trains usually still must stop at many stations. For economical practicability, trains usually must stop at intermediate stations in order to serve as many manufacturers of commercial products and material or individual personal travelers as possible.

The power of this electromagnetic technology needs an atmosphere to have the ability to work. Spacecraft could use an electromagnetic propulsion slingshot method to get it out of the atmosphere of the Earth. Once out of the earth's atmosphere, nuclear energy could supply the required energy to drive a ramjet to go to far distances to remote planets in our solar system. Following is an article describing a slingshot method using electromagnetic propulsion to get a spacecraft into Earth's orbit.

In 2005, NASA engineers created its first electromagnetic propulsion system using a set of devices consisting of a linear motor and a ramjet engine. It would represent the first time that a

ELECTROMAGNETICS

ramjet could take flight at speeds exceeding the sound barrier. Since about 1943, very little progress occurred. In the late 1990s, NASA decided to use a linear motor and a ramjet engine propelled using electromagnetic propulsion rather than having rockets propelled by using liquid or solid base fuels. Delivery of satellites into orbit will occur using this linear/ramjet method instead of using polluting fuels. Much more cargo space can take the place previously needed for fuel. This also should lead to more and better technology to allow greater research in exploration of our solar system and possibly beyond.

The linear motor and ramjet system works basically by first placing the space vehicle or satellite on a long track that would eventually face upward. Then by using electromagnetic propulsion, these objects will very rapidly reach speeds that will place the space vehicle or satellite into orbit around the Earth. The process will help objects like satellites and space vehicles get past the sound barrier and basically hurl the objects into space orbit. NASA further indicates that someday this technology could be used by airplanes and automobiles.[39]

The article, "How Electromagnetic Propulsion Will Work Beyond Our Solar System," shows the possibility of using a nuclear reactor to get to far reaches of our solar system and possible a little beyond.

A nuclear reactor could power an electromagnetic-propulsion system and the spacecraft itself, which would allow space vehicles to travel to the far reaches of our solar system and beyond. As an analogy, powering a space vehicle would find similarity to the powering of a nuclear submarine. One pound of uranium about the size of a baseball provides an amount of energy equivalent to about a million gallons of gasoline. This would allow space vehicles to have the ability to travel incredibly long distances for extended periods of time. These space vehicles would not have an ability to reach the closest star, but perhaps these space vehicles could reach speeds of about 1 percent of the speed of light (186,000 miles per second) or about 6,696,000 miles per hour. NASA admits that there still remain many issues for scientists to unravel, but the concept holds much promise. Keep in mind that as much as a

---
39   "NASA Creates Electromagnetic Propulsion System Prototype," Tudor Vieru.

century ago, people mostly believed that there existed little or no possibility of ever reaching outer space at all.[40]

The previous article discussed what is possible in some future time. Add to that the problem of landing on a far away planet. And add to that, if the planet has a gravity pull anywhere near that of the earth, how can the spacecraft ever get back into space in order to leave that planet and return to earth? Possibly astronauts could build a space station for scientific purposes and for storage of parts and material for eventual use on this faraway planet. After getting enough components on this faraway planet astronauts could build a launch system adequate for hurling an object (a relatively small) spacecraft into orbit. The relatively small orbiting spacecraft would have the ability to reach a longer range spacecraft in the same orbit. The larger space vehicle could then leave orbit and return to Earth.

Admittedly, if this would prove possible and practical, its implementation would not likely happen for quite some time. However, NASA is working on this type of future space exploration. NASA does work on very many futuristic space related projects. They have quite talented scientists who do the work while they also develop many new theories. NASA often gets criticized for spending so much, but I believe, even with some waste NASA is a valuable American asset. We need to be proud of NASA.

Rumors abound that much of NASA's operations may become privatized to reduce costs. If this happens, companies likely would quickly go into the business of selling launches and development of space stations to NASA. While this appears to be adding a middleman, lobbyists will probably continue to pursue the concept.

The main purpose of this chapter consisted of some ancient history of magnets and the long ago accidental discovery of Electromagnetics by Dr Gregory and his description of electromagnetics in 1820. The latter part of the chapter intended to give some idea of the awesome power of electromagnetics.

---

40 *"How Electromagnetic Propulsion Will Work,"* http://science.howstuffworks.com/electromagnetic-propulsion2.htm

Maglev trains, they go so fast.
Maglev trucks and automobiles will happen; the die has been cast.

Chapter 29:
Maglev Train Electromagnetic Levitation

There are currently three electromagnetic levitation and propulsion methods used by maglev trains: Electromagnetic Suspension System (EMS), Electrodynamic Suspension System (EDS), and Inductrack, a newer variation of EDS.

The EDS system, developed by the Japanese, uses a repelling force of magnets and super-cooled, superconducting electromagnetics. Their system allows the coils to continue supplying electricity even after the power supply shuts down. The Japanese electromagnetic system saves energy but super cooling can prove expensive.

The maglev train levitates almost four inches above the roadway. When not traveling more than about 60 mph, the train must travel on rubber wheels. The Japanese indicate that this as an advantage since the train could still operate on the rubber wheels using battery power albeit at a very low speed. The duration will last for a limited time and it depends on the availability and size of the train's backup power supply. A negative issue, the maglev train will need shielding equipment if persons with pacemakers travel on it.

Inductrack is a better, more recent variation of EDS that uses permanent room-temperature magnets to produce magnetic fields. This process uses a power source to accelerate the train until it has enough speed for the system to cause it to levitate. Should the power source fail, the train will gradually slow down and eventually stop on its auxiliary wheels. Engineers have developed two versions of Inductrack: Inductrack I and Inductrack II. Using

Inductrack I, vehicles will operate at high speed. While using Inductrack II, they will operate at lower speeds. Scientists developed Inductrack II to satisfy safety and cost concerns. As long as Inductrack maglev trains can travel as much as a few mph, the train will levitate slightly less than one inch above the roadway. While there is no one indicated as authoring the article, Dr. Richard Post with Livermore National Laboratory, came up with the Inductrack system. Dr. Post came to the attention of NASA, which awarded him a contract to explore the idea of using the Inductrack process to send satellites into orbit.

There has been one accident due to a fire allegedly caused by an electrical problem. No injuries resulted. A second accident happened due to a repair car being left on the roadway. Most of the twenty-nine passengers on the train were killed. No known accidents have occurred due to a failure actually caused by a malfunction of any maglev train.[41]

---

41   "How Maglev Trains Work," http://www.howstuffworks.com

*We must know safety conditions, routing, statistics, and simulation.
Gathering of relative data will occur; it's a natural expectation.*

Chapter 30:
Data Entry, Gathering, and Its Use

Computer systems, using sensors to control and monitor vehicles, will look for the existence of roadway or vehicular problems; this will ensure safety while on the HEVER system. Gathering of data relating to safety conditions, routing, statistics, and simulation will happen naturally while the vehicle, electrically powered, will travel while on the HEVER system. Prior to a traveler's vehicle entering onto the HEVER system, the vehicle will use the central computer system and its own computers to check the condition of the vehicle itself. Assuming acceptable conditions for the vehicle to enter and use the HEVER system (this includes checking for weight limits), the traveler will have the ability to give instructions using a keyboard, touch screen, or voice input device and a monitor on or near the console (dashboard) of the vehicle.

To make the journey as automatic as possible, one will use the vehicle's data entry system to either enter the data about the location of the destination directly into the vehicle's computer system or onto a card that can be used repetitively. Once the destination data is into the database of the vehicle's computer system, the data can be recalled for repetitively (possibly obviating the creation and reuse of the card). Someone can use the card to enter the data into the vehicle's computers before the vehicle enters the HEVER system and, therefore, before starting the journey. Any data entry done (on a card) and double checked prior to going on a journey will help to reduce errors such as wrong destinations. Data (keyed or from a card) entered into the computer's database system will then cause the start of a journey. Travelers can make destination changes by using the vehicle's data entry system to enter any desired change while in transit. The vehicle will display the traveler's requested destination along with any other available alternate destinations along the paths of the original and of the newly requested destination. The traveler must reaffirm his choice, change the destination request, or cancel his request. If the traveler reaffirms his choice it will cause the vehicle's computer system to automatically execute valid changes requested and reaffirmed by the traveler. In a cooperative fashion, computers in the HEVER system's central computer system and computers located in the vehicle will act to carry out the traveler's instructions, sending the vehicle toward the desired destination.

The vehicle's computer will automatically collect and send statistics such as billing data, charges for electricity, parts age, wear statistics, mileage, and efficiency of motors to the central HEVER computer system. The HEVER system will, as appropriate, keep the collected data in the vehicle's and central system's computer database. On a real time basis, the vehicle will provide off-line or on-line information about the vehicle's level of ability to continue to operate.

The data from the cards (like a credit card) will automate determination of trip's destination and the automatic use of routings, unless the traveler has entered different routings. The central computers system will keep data relating the number of times the traveler requested them, changes requested, and routings used.

Manufacturers will have computer terminals at factory or warehouse locations that will communicate with the HEVER system's central computer. Personnel at any entry locations will have the ability to create destination information on cards that will look like credit cards, at any time prior to their use by vehicle operators when they decide to use the HEVER system. The cards will contain all needed trip information.

If traffic volume or other transit problems cause the HEVER system to recommend use of an alternate routing, its computer systems will advise as to which alternate routings to take and how long and what distance the vehicle will travel. If necessary, the HEVER system will even force taking an alternate routing or leaving the HEVER system. If using alternate routings causes a vehicle

DATA ENTRY, GATHERING, AND ITS USE

to leave the HEVER system, travelers will, thereby, loose much of the automating of the trip because of a necessity or desire to use roadways other than the HEVER system. Adjustments will cause changes only if a refund is in order. A trip was calculated based on the final destination and changes in routings will not cause increase in charges unless the traveler requests changes. If the requested change causes an increase in charges, the system will only allow changes if the traveler does not attempt to use cash, unless the vehicle traveling on HEVER system has equipment to allow direct cash entry. When the traveler uses cash at a kiosk where his vehicle enters the HEVER system, his vehicle's computer system will advise him or her as to whether the driver of the vehicle has enough cash to complete the trip. If available cash cannot yield an ability to complete the trip, the vehicle's computer system will reject the routing change request.

While in his/her vehicle, traveling on the HEVER system, the traveler can use time for activities other than driving during any normal trip. When not on the HEVER system, the driver will need to devote virtually all of his/her time to driving. Computer systems of both the vehicle and the HEVER system will provide control of the vehicle and the monitoring of the vehicle's system and components. These computer systems will gather diagnostic information and other statistics and they will have the ability to provide some (non-Internet based) entertainment to the passengers.

The HEVER system's central computer system and the vehicle's computer systems will combine to perform checks of the vehicle when it attempts to enter the HEVER system. When the vehicle attempts to enter the HEVER system, its computer or the HEVER system's central computer system will prevent vehicles from entering if the diagnostics indicate that the vehicle has existing or potential electronic, sensor, mechanical, or computer problems.

While traveling on the HEVER system, the traveler can know the distance left to the destination, locations of any of the oncoming exits in their path, where restaurants serve food, and where they can use available restrooms. Travelers will have access to computer systems usable for entertainment or work while traveling. If the traveler has access to wifi, he/she can use his/her own personal computer to do personal communication, for example, or he/she can make or cancel hotel reservations while the trip is ongoing.

Anything that travelers will need to know about the trip, from starting point to reaching the destination, they can know it by using the information obtained from the computers that essentially control the vehicle. The system will not allow use of the HEVER system's communications capability to contact the world outside of vehicles' computer systems or the HEVER central computer systems. The HEVER system computer system and vehicles' computers must have a closed communication network.

The HEVER system will automatically do the billing and charge the trip cost to the owner of the vehicle unless some traveler other than the vehicle owner provides a credit card (or cash) to use for charging. Communication between the vehicle's computer system and the HEVER system's central computer system will easily accomplish accounting for electric power usage. Each vehicle's computer system will have its own unique identification code(s) that the central computer system can identify. The vehicle's computer system will do the billing in a relatively straightforward and simple manner, much like that for a telephone or credit card bill. The HEVER system will handle the collection of credit card charges and relay them to the appropriate credit companies.

*Repairs, maintenance—looks like an alternate route will loom.
Let's get off of the Interstate and get a motel room.*

Chapter 31:
Construction and Maintenance - HEVER

Construction of the HEVER system needs to guarantee safe, efficient, and fast travel. Since construction personnel will build the system safely above ground employing elevated roadways, the new roadway itself will enhance delivery of material for continuing the construction of the rest of the roadway. After fully implementing parts of the HEVER system, maintenance vehicles can ride on the roadway to the place requiring maintenance or repair. If executing the maintenance or repair takes a portion of the HEVER system out of service, vehicles will likely need to use alternate routing. In these cases of executing repairs or handling some maintenance situations and where travelers would need to use undesirable or impractical alternate routing, a portable, temporary roadway will allow vehicles to go around the trouble spot until fixed. This assumes feasibility and practicality in which a temporary roadway, most likely using interchangeable parts, could quickly be placed to go around problems that would otherwise prevent using a significant part of the HEVER system.

Collision avoidance needs hardware and software for a safety solution. A computer solution will need to control electromagnetic propulsion.

Chapter 32:
Inertial Forces

Today, the need to reduce truck overweight violations has resulted in the creation of a very difficult and costly weight management system now necessary to ensure trucking and other vehicular safety. Inertial forces mostly predicated on weight and speed, determine the minimum distance needed to stop and maximum distance needed to get up to the proper speed. We call the minimum time needed to travel the distance between vehicles "headway." These inertial forces dictate power requirements for getting loads up to and maintaining required speeds especially for merging into the traffic mainstream.

The HEVER system computer and the vehicle's computer will jointly manage merging as necessary (collision avoidance). Design, using electromagnetic reverse propulsion, will adequately satisfy stopping requirements of braking system; the friction brake system will not find use while vehicles use the HEVER system.

Due to the requirement to accommodate inertial forces, and to maintain proper headway distances between vehicles, the HEVER system will slow or stop any vehicles as required. On an automated transport system such as the HEVER system, safety requirements will absolutely demand that appropriate design personnel establish proper weight maximums and provide an overweight detection system that will guarantee safety. The HEVER system central computer and the system on the transporting vehicle will have software and hardware that will do a more than adequate job of ensuring that vehicle weights are within allowed weight maximums. The HEVER system central computer will

HEVER

discover and prevent truck drivers with overloaded vehicles from entering the HEVER system roadway. The vehicle's computer system will assist the prevention of overloading when the company loads the vehicle, long before ever attempting to enter the HEVER system.

    The HEVER system will likely require the setting of lower or higher load maximums than those in use today. Designers will set weight maximums needed to avoid undue wear and tear to any existing roadway used by vehicles when not traveling on the HEVER system roadway. Roadways of the HEVER system above ground will have an additional requirement to maintain appropriate maximum weight standards on the roadway. A safe system such as the elevated HEVER system roadway must not and will not allow too many vehicles, too close to each other or vehicles weighing too much. Designers will control the inertia issue and not let it become a problem by designing the HEVER system roadway strong enough and by monitoring weight allowances under control of the central computer system of the HEVER system. The roadway of the HEVER system must allow and handle a high volume of heavy loads. The HEVER system must allow ease of automatic and manual (visual?) checking of trucks' freight containers just before they try to enter the HEVER system.

*When leaving the HEVER system do you
need a temporary parking spot?
If so, let the computer do it for you; that's hot.*

Chapter 33:
Parking Vehicles Leaving HEVER

Since the HEVER system design will allow automated movement of unmanned vehicles, the system must allow for enough exit space to bring transport vehicles safely from the HEVER system. An adequate amount of rail distance must exist for the purpose of slowing the vehicles that are leaving the HEVER system. When leaving the HEVER system, most drivers of vehicles will leave it without parking and will finish the trip by driving on to their final destination. If necessary to park (for example, an emergency such as unavailability of a driver, e.g., unconsciousness), there must be adequate space for vehicles to automatically park at the destination. The vehicle's computer control system will accomplish this by using a software program that will make the vehicle follow a path to a temporary parking spot. A high charge will be exacted for using parking spots for convenience. Not much time should pass before towing a car to a remote parking facility.

The trip's original drivers, normally in the vehicle and available, will operate the vehicles when exiting the HEVER system. If no one occupies the vehicle or if the vehicle has an incapacitated operator the vehicle will require a temporary parking spot for the queuing of their vehicles as it leaves the HEVER system. If the HEVER system's central computer determines that the temporary parking area for vehicles has filled too close to full capacity, some vehicles that have been parked the longest will automatically travel to the closest overload parking facility on a first in first out basis. The HEVER system will exact a relatively high relocation charge. Normally there will be adequate parking space, but some vehicles in this sort of overflow parking situation may need manual removal from that parking area and traffic control authorities will then need to accept them into their purview. People should not temporarily or permanently abandon their automobiles in HEVER system parking areas. However, if they do, the situation will eventually cause the manual removal of vehicles. This emergency procedure for manual removal of vehicles should not occur often since drivers will normally reside in their vehicles and will completely leave the HEVER system upon arrival at the requested exit.

Trucks will sometimes arrive without a driver and will require a parking facility most likely separate from that for automobiles. The parking facility for trucks will need to be a larger and more highly managed parking area than that for automobiles in order to allow adequate queuing of trucks leaving the HEVER system. Most trucks will likely not require this temporary parking. Some trucks will have drivers either already in the truck or waiting to drive the truck if it parked automatically. Still others will exit at spurs used by manufacturers, distributers or other mass spur users.

Prior to leaving the HEVER system on a spur, if a parking facility at a manufacturing or warehouse destination overflows and the truck driver's vehicle would not immediately exit the HEVER system at the spur, the truck will automatically travel to the closest general parking facility for short term temporary parking. Once available parking space exists at the factory or warehouse's spur exit, the delayed vehicle will be moved from short-term temporary parking and will be sent on its way.

If short term parking time limit has not expired causing manual removal to a non-HEVER parking facility, this provides alternate locations for temporary, somewhat emergency parking for vehicles until they can proceed to their exit. With this type of store and forward capability, the HEVER system can accomplish the transporting of freight by getting the vehicle and its contents as close to the destination as possible in the event that the transporting vehicle cannot get all the way to its intended destination. These overflow parking facilities will not allow any long-term parking. They must always maintain (guarantee) a high availability of parking area for short-term parking.

*So many vehicles will face ultimate replacement.
While taking long trips, HEVER vehicles' tires
will travel on less cement.*

Chapter 34:
Some Notes on Future and Change

Nominally, about 250 million or so operational automobiles, trucks, and buses operate on roadways in the United States. Bringing the HEVER system up into full operation that will eventually require an evolutionary replacement of most of these vehicles. Needless to say, this will take many years, but we should build the HEVER system soon and get started on replacing our older vehicles with new electric powered, HEVER capable vehicles. While this will cause some displacement of workers, full development will take a lot of time. In the long run, this will lead to better jobs, better travel and we will burn less oil; we will reduce pollution. HEVER system development will require institutions producing oil to concentrate on markets where oil usage doesn't involve any significant amount of waste and pollution. We will certainly have enough uses for oil without burning it. Demand for oil will decrease, but this actually good happening will allow oil companies to produce and sell oil (and some new complementing products) for a much longer time.

The GM Chevy Volt automobile has a motor that has an all-electric design It will operate using a gas powered generator to provide electricity to the GM Chevy Volt should the battery become depleted. While this vehicle still burns some oil, an EPA rating gives the Volt a 230 mpg rating using a new city and highway method of rating. At any length, this will help to evolve our country away from its huge appetite for oil.[42]

---

42 "Chevy Volt Gets 230 MPG City EPA Rating," August 11th, 2009, http://gm-volt.com/2009/08/11/chevy-volt-gets-230-mpg-city-epa-rating/#.

HEVER

Two or more HEVER central computer system(s) might actually locate regionally and if each system contained all of the same data, they could back each other up as parallel real time operating computer systems. Each vehicle will need many functionally specialized (very low cost) computers and sensors. Considerable computer and other vehicle components will need to exist, sometimes multiples doing the same task, for backup and for using a voting process in case of component failure. This will bring a boom to the automotive, computer hardware and computer software industries. Assume that automobiles, before adding in the cost of computer equipment, could be made and sold for an average of $20,000. Automobile sales in the United States alone will reach $6 trillion. The total replacement cost of the $6 trillion would double to $12 trillion if the individual average vehicle costs turn out to be $40,000 (double the $20,000). If computer hardware and software costs add $3,000 per vehicle, semiconductor manufacturers of computers and software manufacturers might sell about $2 trillion worth of product. Add in a few more needs or wants and we will be getting close to or even exceeding a cost, over the many years of development, of $15 or $20 trillion. This could all be developed in the US.

This will likely take at least ten years to get established and over fifteen to twenty-five years for the HEVER system to mature, so we had better get going.

Just look at the explosive Internet growth and you can observe today's hot information and communication industries. Compare the value of business oriented communications and information technology to our need for improving individual mobility through the redesign our nation's transport systems. Difficult to over rate the value of the development of the HEVER system, it will save lives, reduce injuries and save time.

Needing to change, airports have become cluttered, inefficient, overloaded, and various functions and services often get out of control. One constantly hears stories of lack of upgrading equipment or of overworking employees. Interstate roadways and other highways have become increasingly overloaded. Railroads can only handle the lowest cost freight that trucks can't or won't

SOME NOTES ON FUTURE AND CHANGE

handle. Light rail passenger services are sophisticated streetcars, a technology that is at least a hundred years old. Bus systems could offer a viable means of mass transit, but they often make too many stops, most of them pollute and more important, from a practical point of view, the masses don't seem to want to use them; people don't seem to want to let go of the use of their individual personal vehicles. Mostly due to the constant growth of high traffic volume, the moving of people and material via the Interstate has changed the Interstate into a virtual kludge. Our current transport vehicles and systems have increasingly grown wasteful of time. They have become more expensive, overloaded, and more dangerous and, as such, extremely outmoded.

If needed due to unusual snow and ice condition on the HEVER System heat tracing might be needed. It will likely prove difficult to determine where it can predictably find need. If designers determine that too large of an area must employ heat tracing, it will likely not be done. The practicability and economics of using heat tracing must be addressed; it may or may not prove impractical or too costly.

Snow and ice conditions on roads and highways today are extremely difficult to handle; the number of accidents increases during bad weather especially during snow and ice situations. The HEVER system is a controlled roadway. Heat tracing in areas of extreme cold weather can prevent icing during inclement weather on the HEVER system's roadways. Accumulation of snow will normally not impede travel because the HEVER system is elevated. Even in very snowy regions, it will be easier to keep the roadway clear. Rain, snow, or icy weather will not hamper the braking system on the HEVER system. In virtually any emergency caused by weather, the electromagnetic propulsion has the capability to move vehicles; it will also have the ability stop the vehicles, immediately if necessary.

The HEVER system is a project with virtually no competitor larger. We will need intelligent professionals and a talented project manager.

Chapter 35:
Organization for Design/Implementation

A development and operation of a large system like the HEVER system will ultimately require enormous man time, material and capital resources (as did the development of our current national Interstate system). A government institution could create an organization to design and implement the HEVER system. A government operation would likely provide the best start for organizing such a large endeavor, but as an alternative, our government might allow the forming of a monopolistic or quasi-monopolistic organization (for example, like a public utility). A top manager, not necessarily technical, should manage the startup and building of an appropriate organization.

As a first step, the organization will need to build a strong managerial team that has great management and technical skills. Management should hire design engineers with strong technical knowledge and leadership skills (shirt sleeve supervisors). They should also hire design architects, with the highest of technical reputations for design of computer controlled systems similar to the future HEVER system. While not limiting personnel search to any given industry, hiring some persons with backgrounds from the railroad, trucking and automotive, and computer hardware and computer software industries might help. Ideally, some of the persons hired should have strong knowledge and experience in statistics, physics, and engineering. Probably the top executive won't necessarily require industry or even technical skills; rather, most important, he will need to have extremely strong organizational skills and the ability to find, attract, retain, and organize managers who in turn can build specialized teams who will design, build, and implement the HEVER system. Wouldn't it be nice if this now organization could steal some personnel from NASA or nationalize (just kidding) the most advanced company currently involved in maglev design and development?

The top management group would need to hire a group of engineers to support management directives and to guide development of design. The early (and later development) design period will require depth of knowledge and experience in statistics. The new organization's engineers will develop enough of a design to allow a bidding process to happen. In parallel to this initial design, they will also design and implement simulation models and produce a requirements document.

Much is known about electromagnetic levitation and propulsion. However, if even possible, changing the application of maglev train technology to maglev truck and automobile technology will prove to present a significant challenge. Much engineering design will need to happen. Development of computer system hardware and software will prove to be one of the largest, if not the greatest, challenge.

Many contract proposals will require study and will result in contract awards. Processes of Request for Information (RFI) and Request for Proposals (RFP) will need to become an integral part of everyday efforts.

In order for potential responders to bid on an RFI, they will need enough information in the request they receive about the organization's planned set of requirements along with some elements of a skeleton design. The RFI should happen prior to an RFP in order to obtain responding companies' information about their management ability, their customers' opinions, product availability, technology, and experience prior to the issuance of an RFP. Don't trust oral promises; require hard copy about their products and services. If responding companies do not market proposed products and services through use of printed manuals and brochures, the products and services may not be real. They may only exist as untested ideas or plans. Of very great importance, responders must provide proof of strong financial viability,

ORGANIZATION FOR DESIGN/IMPLEMENTATION

ability to survive and have enough expert personnel with ability to handle projects upon which they will bid.

Responses to both an RFI and an RFP should give a complete picture of the capabilities of potential vendors. Their responses should relate to accomplishment of design and implementation of issues like those stated in the RFI. A company response to an RFI should relate to the tasks to be done. It should give very detailed data about the responding companies.

RFI response documents must become the property of the organization (initially possibly a governmental group). The managing organization may want to allow RFI responding companies a limited monetary grant to help encourage better quality RFI submission. Depending on vendor management decision, a limited grant dollar amount might encourage responders to agree to respond to an RFP or not. It should help responding companies immensely. The future winner would likely receive a lot of contract work in the future.

Because an RFP basically contains a refinement of the responding companies' belief that they can satisfy the organization's requirements stated in the RFP, a response to an RFP is similar and should actually provide a refinement of responses to an RFI. The RFP, actually a requirements document, will consist of the functional needs of the system in increasingly greater detail without limiting respondents' planned implementations to specific techniques, methods, materials, products, or vendors. After completion by responders, the responses to the RFP will, to some degree, mirror the requirements document and provide an overall design of the system. Responding companies may present new ideas not realized by the creators of the requirements document. Not a perfect document, change to the requirements document must happen if appropriate. As with the RFI response, all responses to an RFP must become the property of the organization. Also, like suggested for the RFI, the organization may want to allow RFP responding companies a limited monetary grant to help encourage higher quality RFP response submissions. Winner of this first phase of work can, upon approval, use any part of any RFI or RFP responses.

HEVER

When issuing the RFP, it will include the requirements, and legal rules to which responding companies must adhere. The RFP will contain a request for responding companies' commitments such as estimated cost and schedules. Responders can spend as much money and man time effort on responses to the RFP since each will want to provide a winning proposal to do the engineering design (probably leading to many years of work). The initial design effort, while not a complete design, should include a proposed design for enough of the system to determine feasibility to go forward with the rest of the design.

RFP responders who submit proposals to win a contract to do the ongoing engineering might receive a limited payment for doing such a large proposal. Fixed amounts of $100,000 or $500,000 (or whatever) should be offered to responding companies who choose to bid. The organization will not pay anything above the original same payment amount given to each company responding to the RFI or RFP. The organization would have given the payments with the intention of encouraging vendors to respond to the RFI and RFP. All potential RFP responders should receive the same dollar amount as that offered to any of the other RFP responders. Perhaps after reviewing responses to the RFI, the three (or at most four) RFP potential responders would be selected for the next bidding step after review of candidate viability. RFP responders can spend more than the fixed amount they are paid to encourage higher quality bids, but if the organization decides to grant funds to potential responders, a limit must be established for paying the potential RFP responders. Possibly something like an initial implementation of four or five cities can be the end product of a fairly detailed design. The purpose of this effort will be to gain more knowledge of how to build and expand the HEVER system. All RFP responders' proposed designs should become the property of the organization.

After choosing a winning company that proposed the "best" design features, that company can use any parts of other responders' proposed designs. As a result of the review of the designs submitted by responding companies, the winning company will continue and the next design round can continue. The next

ORGANIZATION FOR DESIGN/IMPLEMENTATION

design should include a specification document. It will consist of great detail as to what the winning company and his suppliers will do and supply; it will contain specifications for quality, types of components, type of materials for vehicles, computers, roadways, and supports. During this phase of (mostly if not all) engineering effort for roadway construction, central computer system, vehicle computer, test facilities, test prototypes, and test scripts will need to happen. On an ongoing basis, the simulation model(s) must be upgraded to include design information as it becomes available.

The design team must be a permanent group extending the HEVER system through the years. When the production implemented HEVER system operates every day, a management and engineering group will take on the responsibility for operation and maintenance of the system.

Can we do without costly computers to run our lives?
If we didn't have computers we would unhappily have the hives.

Chapter 36:
Computer Systems

Possibly all vehicle's computer systems and kiosks' computer systems could provide virtually all necessary (complete) computer control. Vehicle's computer systems would typically provide data gathering, store and forward functions, receive of software upgrade downloads from kiosks and, monitor, and control what is done by vehicles' computer systems. Minimally, a central computer system will be needed for downloading software upgrades to kiosk computer systems and receiving data gathered by vehicles' computers (relayed to kiosks) from kiosk computer systems. Kiosks' computer systems could receive the data gathered by vehicles' computer systems either in real time or in a batch when the vehicles leave the roadway, whichever is more practical or necessary.

In many ways, it appears that the HEVER system might need less of a central controlling computer system, especially if its reduction in function allows a design that reduces costs without negatively impacting safety. A large, totally integrated central computer system (maybe even a super computer) would certainly add capability to the overall HEVER system. However, the relatively small computer systems in each vehicle and at each computer system at entry and exit kiosk locations could provide the necessary functions. If we have a smaller central system, having it must not create negative safety issues. The central system would exist basically to integrate everything, thereby bundling some accounting functions and taking some control capability from the vehicles' computer system. It may prove more economical to have kiosk computer systems handle more of the temporary storage and relay appropriate data gathering to the central system for use in testing and simulation.

The computer system needed for testing and simulation could also provide a facility for handling software development and software delivery to kiosk computers, which would in turn load the software onto vehicles when they use the kiosk. Also anytime desirable, owners could go to kiosks to allow downloading of software upgrades to their vehicle's computer system. Some of the computer software upgrades downloaded to vehicle computer systems might also provide capabilities for use while traveling off the HEVER system. The computer system used for testing and simulation could also have the necessary software to handle the organization's administrative functions.

Kiosk computers could more directly send charges, such as credit card fees, to appropriate destinations.

Assuming a smaller rather than larger central controlling computer system, a differing amount and type of computer software would need to exist in vehicle computers. Having more software function in kiosk and vehicle computer systems probably would allow easier development and upgrade of software for them. When a vehicle needs to enter the roadway, the kiosk's and the vehicle's computers would need to communicate with one or more vehicles' computers immediately coming toward the entry point in order to know when enough headway exists to allow safe entry onto the roadway. Kiosk computers would need to have the ability to communicate with the one or more immediate vehicles coming its way to maintain control of the vehicle leaving the kiosk's exit. The vehicle and the kiosk computers would need to safely park the exiting vehicle if necessary. The kiosk computer would need to know that adequate parking is available if needed.

During the initial engineering phase, designers will need to make a final determination of the amount of central computer power to have and whether or not to provide a more highly integrated central computer system. If designers choose to provide a relatively small central controlling computer system, the HEVER computer system would still need a communication system. It would likely consist mainly of cabling; using hardwired cabling (rather than wifi) to communicate between vehicles and entry or exit kiosks. A shielded hardwired cable will provide more reliability than will wifi; it will also provide more security from hacking. A sizeable computer system will need to exist at the facility for software development, testing, and simulation purposes. Again, designers will need to make a final determination as to functions to assign to kiosks' computer systems and functions to assign to the software development test and simulation (central?) computer system (or another computer system?) for sending the required data where needed in real time or in a batch mode at scheduled times.

Sir, I have deep concern that your project will badly turn.
Oh yeah! Let me say, "The road to hell is paved with deep concern".

Epilogue:
Reasons for Writing a Book

I have always concerned myself with man's penchant for waste and pollution. We shouldn't even throw our personal trash out of vehicle windows as we drive on the nation's roadways. In our nation's history, we have often done quite well (sometimes barely) at gaining and preserving our freedoms. However, so many times our country (and often the entire planet) has experienced loss of opportunities for making unusually great progress, mostly due to powerful special interests and fear of the unknown. I have heard someone say, "We may not do the best that we can do, but we continue as the best, most advanced country in the world. We have the best standard of living and everyone wants to live here." Sadly, I'm not so sure that gives a true picture, and even if it does, for how long? Russia and Germany probably have superior mathematics and physics universities than we do. China has become an industrial giant. India has developed into a computer software powerhouse. The Japanese and Koreans have the reputation of building cars that have extremely good quality and reliability; they are also leading manufacturers of electronics.

Our planet currently has far less magnetism than it did a century or two ago. Has our planet reached a point of becoming a dying planet? I believe that whether global warming happens or whether it does not happen to the Earth, we need to have more respect for keeping it clean enough for our children and their children. We actually should leave the Earth better than when we occupied it. There exists a possibility that a long-term cycle might bring us back out of the global warming. Regardless, we shouldn't pollute our planet and at the same time waste resources such as oil. Even if oil didn't pollute, we shouldn't burn this irreplaceable resource. We must save oil for other uses and other generations. Currently, even the use of coal has resulted in a significant pollution problem. If global warming becomes absolutely proven (and if so proven, I hope at not too late of a time) perhaps we should dredge earth from the seas and use the earth to build up our shores somewhat like the Netherlands did. Also we might consider the building of more dams to preserve fresh water.

For at least twenty years, I have kept notes about and have wanted to write this book about mass transit. Of course, until recently, the concept of electromagnetic levitation and propulsion didn't really exist as a possibility; it was mostly a theory. In the book, I talk about maglev trains as wrong for America and maglev automobiles, trucks, and buses as right for America.

Maglev means magnetic levitation using electromagnetic propulsion. I have discussed this more in the book.

- Many people fear spending money, especially for big projects; a few people fear not spending.
- Maglev trains – established concept; maglev automobiles – very new
- Maglev trains- high speed accepted; maglev automobile, truck, and bus high speed – fearsome; note that some sport cars currently have the ability to travel very fast, about 250 mph.
- Maglev trains – high capital investment and operational cost; maglev automobile, truck, and bus – high capital investment and operational cost. Automobiles, trucks, and buses should prove more cost effective and provide greater benefits. Through charging for use of the HEVER system, we should be able to return the capital investment costs. Ownership cost will likely provide a better long-term value than that for maglev trains.

Please keep in mind that trains obviated the development of canals. Trucks took away much of the train market. Maglev cars, trucks, and buses are relatively non-stop in that you don't have to change vehicles from the beginning of the trip to its end. There currently exists quite a strong movement to build computer controlled maglev train systems. If this happens, in the long run, cars, trucks, and buses will again take away any perceived maglev train use advantages. They will even take away much transport done by the current train industry, such as piggyback transport of cars and truck trailers. Computer control of maglev cars, trucks, and buses will save lives by virtually eliminating accidents. So, are there any reasons why anyone should read this book and give the concept a chance?

- Long-term ownership costs favor maglev cars, trucks, and buses over maglev trains.
- Long-term benefits favor maglev cars, trucks, and buses over maglev trains.
- Maglev cars, trucks, and buses can travel non-stop; maglev trains makes stops. Economics for maglev trains will require stopping at stations to pickup or drop off material and people. Making more stops is one way that trains make some of their revenue.
- Maglev cars, trucks, and buses pick up material at sources away from the hi-speed roadway; maglev trains do not.
- Maglev cars, trucks, and buses deliver contents to final destinations; maglev trains do not.
- Maglev cars and trucks will require no baggage checking; maglev buses will require minimal baggage checking; maglev trains require more baggage checking and baggage retrieval at destinations.
- Maglev cars and trucks will not require parking your vehicle when starting trips; maglev buses and maglev trains require more automobile parking space.
- Maglev cars and trucks will not require renting vehicles when reaching ending trip destinations; maglev buses and maglev trains will require vehicle rental or other modes of transportation at ends of trips.
- Maglev cars, trucks, and buses will normally yield better travel experiences than those provided by maglev trains.

I wrote this book knowing quite well that it is an imperfect, conceptually unfinished document. One only has so much time and then an idea gets old and sometimes becomes infeasible due to other competing ideas growing into institutions. A significant reason that ideas die before they even get studied has to do with other competing ideas. For example, the maglev train represents a competing concept to the ideas I have presented. Both concepts would like to have the capital funds and it will be more difficult, if not impossible, to do both.

The maglev train concept seems to promote an idea that if you give society mass transit, serving the group of users, people and commercial businesses will want to use it even more than owning and using individual vehicles.

I believe that many people want other people to use group mass transit so that they can drive there individual vehicles on less traveled roadways. The HEVER system will make its appearance as a truly mass transit system. It differs from most past mass transit concepts in that it transports individuals or groups in individual vehicles.

In the mid-1950s, St. Louis had one of the best mass transit bus systems in the country. Most people gravitated to automobiles; it was an easy sell for the automotive industry. Making a greater volume of cars helped to cause the replacement of the comparatively few buses. A similar phenomenon occurred where the truck industry took much of the train industry's market.

There are pockets where mass transit buses and trains are highly used for personal travel—places like New York. But even in New York, the city is filled with taxi cabs. I believe that someday people will travel by computer controlled vehicles into the heart of New York City and the system will use streets elevated above the ground. Developers will provide inexpensive, highly available, highly automated, and mostly free parking.

There will, in the foreseeable future, exist situations where specialty non-HEVER capable vehicles (for example off-road vehicles), especially trucks, that will need to have drivers operating them. Even while needing those vehicles, electricity should power them. Since most drivers will want to use the best and fastest roadways, other than the occasional need of specialty vehicles, persons will eventually do very little actual driving of their vehicles. Most travelers will prefer automated transport. In the future, even when

REASONS FOR WRITING A BOOK

persons find it necessary to actually drive their vehicles, computers will significantly monitor driving and do some vehicle control. Maybe if a person couldn't afford an automobile or didn't want to own one, he or she would use mass transit and more than likely buses would provide the service, often using the HEVER system. Otherwise, if you could use your private vehicle to travel so quickly between Philadelphia and New York (about fifteen minutes one way), why would you want to use any other type of mass transit?

Even if mass transit for groups exists in the future, it probably will never again experience its relatively high use as it did years ago. Young people might continue to use some form of "group" mass transit (typically buses) as it exists today when they are too young to drive. However, you only need to look at most high school parking lots. During school, they are full of mostly student driven vehicles. Most young persons can hardly wait until they reach the age permitting them to drive. College and university parking lots are huge and full of student driven vehicles.

Light rail has never gotten to a point of economic success. It doesn't pay for its day to day operation, let alone try to return anything to repay capital development costs. It enjoys the use of subsidies. There is a good treatise on the subject of light rail written in October 1994 in *The Regional Economist*, a document published by the Federal Reserve Bank of St. Louis. I believe the situation of light rail economics discussed in the Federal Reserve document still persists.[43]

People of who have the privilege of driving, normally prefer to use their own private vehicle. They just want to get to destinations quickly, safely, and at a reasonable cost. They also want good, free, or at least low cost parking facilities.

It is the Author's hope that this idea of Hi-speed elevated transportation will be given a chance to occur. If engineered feasible studies are done and the HEVER system is designed and implemented, it will save lives, help the greening movement, reduce time for various types of transport, and will make long distance travel more practical. It has the potential to significantly raise our standard of living.

---

43   Adam M. Zaretsky, "Riding the Rails a Look at Light Rail Transit," The Regional Economist, October 1994, 4.

With each new idea for moving forward, we can decide to let the idea pass, or we can apply engineered study, and try to determine if it should be done. I sincerely believe if we say, "Pass" and don't build a better, highly computerized roadway system, other countries will; and then, they might, more and more often, look at the USA in their rear view mirror.

Bibliography

"A Summary of Important Legislation." UK Department of the Environment, http://www.roadsafetyni.gov.uk/index/road_safety_education/teacherzone-home/teacherzone-mvrus/mvrus-legislation.htm.

Ansari, Azadeh. "Oldest Human Skeleton Offers New Clues to Evolution." CNN, October 7, 2009, http://www.cnn.com.

Berdichevsky, Gene, et al. "The Tesla Roadster Battery System." August 16, 2006, http://www.teslamotors.com/display_ data/TeslaRoadsterBatterySystem.pdf.

"Bob Lutz Implies Chevy Volt Will Get Between 40 and 50 MPG in Charge Sustaining Mode." October 27th, 2009, http://gm-volt.com/2009/10/27/bob-lutz-implies-chevy-volt-will-get-between-40-and-50-mpg-in-charge-sustaining-mode/.

"Bullet Train Launches High-speed Rail Service." *St. Louis Post-Dispatch*, April 4, 2004.

"Chevy Volt Gets 230 MPG City EPA Rating." August 11, 2009, http://gm-volt.com/2009/08/11/chevy-volt-gets-230-mpg-city-epa-rating/#.

Cox, Terry. "Railroad Land Grants." Undated, http://www.coxrail.com/land-grants.htm.

Crouch, Elisa. "Drivers Ignore Hwy. 40 Barriers." *St. Louis Post-Dispatch*, September 2, 2009.

Definition using Merriam-Webster's Online Dictionary, retrieved from www.Google.com/Dictionary.

"Environmental Pathways, Environmental Contamination and Other Hazards." CDC, http://www.atsdr.cdc.gov/HAC/pha/wel_sp/wsta_p2.html.

Faulkner, Harold Underwood. *American Economic History*. New York: Harper & Brothers Publishers, 1949.

Fogel, Robert W. "Railroads and American Economic Growth." In *The Interpretation of American History*. New York: Harper & Row, Publishers, Inc., 1971, 187-203.

Freeman, Sholnn. *Washington Post*, August 23, 2006, http://www.washingtonpost.com/wp-dyn/content/article/2006/08/22/AR2006082201152.html.

Grego, Peter. *The Universe*. London: Harper Collins, 2006.

Hambling, David. "Super Concrete in the U.S. Military, Iran… and the Pyramids?" October 22, 2009, http://www.wired.com/dangerroom/2009/10/super-concrete-in-the-us-military-iran-and-the-pyramids/.

"How Maglev Trains Work." http://www.howstuffworks.com

"How Electromagnetic Propulsion Will Work." http://science.howstuffworks.com/electromagnetic-propulsion2.htm.

"Hollingsworth, Brian and Arthur Cook. *The Great Book of Trains*. Pub location??:Salamander Books Ltd., 2003.

BIBLIOGRAPHY

"IBM Muscles into CIGS Solar-cell Market." May 15, 2008, http://www.physorg.com/news130086323.html.

"In the United States 42,815 People Were Killed in Road Accidents Last Year." *St. Louis Post-Dispatch*, Monday, April 5, 2004.

Kent, Alan, and Micro De Cet. *The Complete Encyclopedia of Locomotives*, Lisse, Netherlands: Rebo Publishers, 2006.

Launius, Roger D. *Space Stations*. Washington:Smithsonian Books, 2003.

Launius, Roger D. and Bertram Ulrich. *NASA & the Exploration of Space*, New York: Stewart, Tabori & Chang, 1998.

Leitman, Seth. *Build Your Own Electric Vehicle*. New York:McGraw Hill, 2009.

Locklin Phd., D. Philip. *Economics of Transportation*. Homewood, IL.: Richard D. Irwin, Inc., 1972.

Mascari, Christopher. "Tesla Model S Sedan Concept: $49,900 Seven-Seater Electric to Hit Streets In 2011." March 26, 2009, http://jalopnik.com/5185844/tesla-model-s-sedan-concept-49900-seven+seater-electric-to-hit-streets-in-2011.

Murph, Darren. "Tesla's Roadster rolls 241 miles on single charge, annoys petrol pumps." April 13, 2009, http://www.engadget.com/2009/04/13/teslas-roadster-rolls-241-miles-on-single-charge-annoys-petrol/.

Vieru, Tudor. "NASA Creates Electromagnetic Propulsion System Prototype." September 19, 2009, google.com/search?sourceid=navclient&ie=UTF-8&rlz=1T4SUNA_enUS25 6US270&q=NASA+Creates+Electromagnetic+Propulsion+System+Prototype.

Ratcliffe, Heather. "$2 Sale of Land Proves Costly." *St. Louis Post-Dispatch*, Monday, 10/18/2009.

Ricker, Thomas. "Tesla Readies New Transmission, Ramping Production." September 10, 2008, http://www.engadget.com/2008/09/10/tesla-readies-new-transmission-ramping-production/.

Savov, Vladislav. "Tesla Roadster Keeps on Rollin', Goes 313 Miles On Single Charge." October 31, 2009, http://www.engadget.com/2009/10/31/tesla-roadster-keeps-on-rollin-goes-313-miles-on-single-charge./

Shepardson, David. "EPA Delays Decision on Boosting Ethanol Blends".*Detroit News Washington Bureau*. EPA Delays Decision on Boosting Ethanol Blends

Landis, Geoffrey A. "Solar Airplane Developed for Venus Exploration." Glenn Research Center, January 21, 2005.

"Some Gas-Guzzling Vehicles Enjoy Sales Surge." *Forbes*, October 23, 2009, http://www.msnbc.msn.com.

Thomas, Lynn. *San Francisco's Cable Cars*, Sparks. Nevada: Lawson Mardon Group, 1992.

Tomich, Jeffrey. "Ethanol is Facing Key Decision Today." *St Louis Post-Dispatch*, December 1, 2009.

U.S. Bureau of Transportation Statistics, http://www.census.gov/compendia/statab/tables/09s1027.xls.

U.S. Bureau of Transportation Statistics, http://www.bts.gov/publications/national_transportation_statistics/html/table_01_01.html.

BIBLIOGRAPHY

U.S. Department of Transportation (US DOT), http://www.bts.gov/publications/national_transportation_statistics/2008/excel/table_04_01.xls.

U.S. Department of Transportation (US DOT), http://www.bts.gov/publications/national_transportation_statistics/2008/excel/table_04_27.xls.

U.S. Department of Transportation (US DOT), http://www.bts.gov/publications/national_transportation_statistics/excel/table_04_28.xls.

U.S. Department of Transportation (US DOT), http://www.bts.gov/publications/national_transportation_statistics/2008/excel/table_04_17.xls.

U.S. Department of Transportation, http://www.bts.gov/publications/national_transportation_statistics/pdf/entire.pdf.

"U.S. Limits Waits on Planes." *St Louis Post-Dispatch*, December 22, 2009.

Waters, Helen. "Why Design Thinking Matters." October 27, 2009, http://www.businessweek.com.

Whipps, Heather. "How the Wright Brothers Changed the World." August 11, 2008, http://www.livescience.com/history/080811-hs-wright-brothers.html.

Woodyard, Chris. "GM Works to Make Some Noise." *USA TODAY*, November 25, 2009.

Zaretsky, Adam M. "Riding the Rails a Look at Light Rail Transit." *The Regional Economist*, October 1994, 4-9.

Made in the USA
Lexington, KY
22 September 2010